SUPER PIPE
The Arctic Pipeline—
World's Greatest Fiasco?

Also by Earle Gray
IMPACT OF OIL (1969)
THE GREAT CANADIAN OIL PATCH (1970)

SUPER PIPE
The Arctic Pipeline—
World's Greatest Fiasco?

by EARLE GRAY

Editorial Consultant:
JAMES T. WILLS

GRIFFIN HOUSE
TORONTO 1979

Published by Griffin Press Limited,
461 King Street West, Toronto M5V 1K7, Canada

Printed and bound in Canada

CONTENTS:

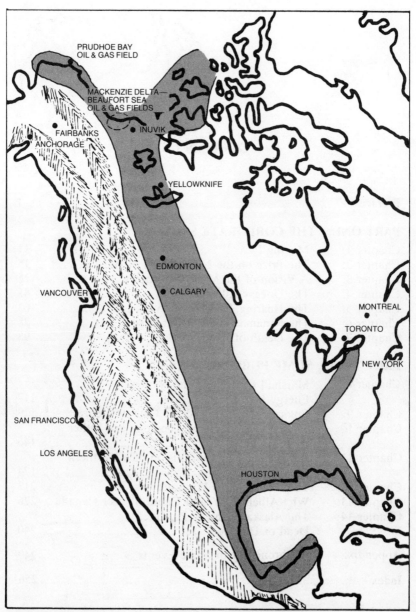

MAP 1. The great interior belt of sedimentary rocks where accumulations of oil and natural gas account for the bulk of North America's petroleum reserves. In the western Arctic, Prudhoe Bay is the largest oil and gas field in North America. To the south, large new reserves of oil and gas have been found offshore from Mexico.

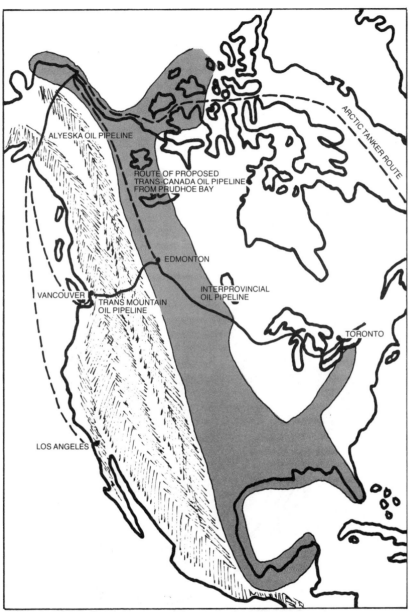

MAP 2. Three alternative ways were considered to transport oil from Prudhoe Bay: by pipeline across Alaska and then by tanker; by pipeline across Canada to Edmonton, connecting with present pipelines; and by tanker shipments through the Northwest Passage.

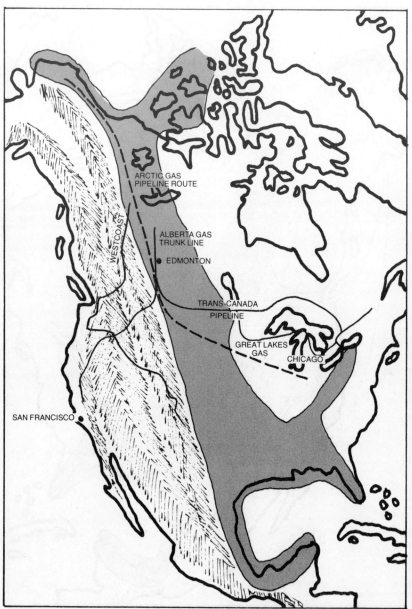

MAP 3. Rejected pipeline route to transport Prudhoe Bay and Mackenzie Delta natural gas along path of sedimentary belt.

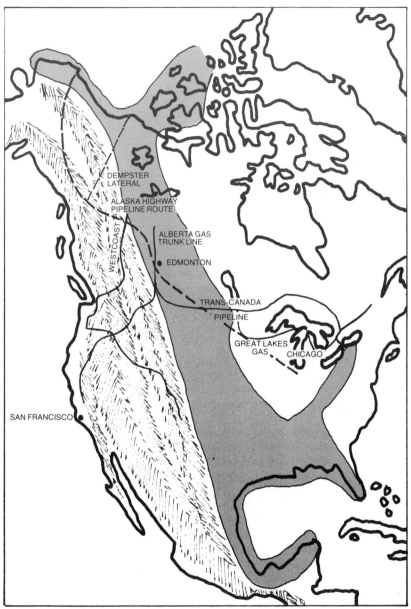

DEMPSTER
LATERAL

ALASKA HIGHWAY
PIPELINE ROUTE

ALBERTA GAS
TRUNK LINE

WESTCOAST

• EDMONTON

TRANS-CANADA

PIPELINE

GREAT LAKES
GAS

CHICAGO

SAN FRANCISCO •

MAP 4. Alaska Highway pipeline route approved by Canadian and U.S. governments to move Prudhoe Bay gas, with possible connecting pipeline link to be built later to transport Mackenzie Delta gas.

Preface

A time bomb is ticking in the Middle East. It could rupture the world's most important source of energy. From the Middle East, from Islamic nations within a thousand-mile radius of the Persian Gulf, comes half the oil used by the non-Communist nations of the world.*

A prolonged, major disruption of Middle East oil supplies would cripple the non-Communist world. Gasoline and fuel oil would be rationed. Energy prices would explode again. Industries would be shut down. Corporations and governments alike would face financial collapse. There would be massive unemployment. Civil strife would be widespread. The poorer nations of the world would endure even greater misery, with increased costs for fuel and food and a cut off of aid from the crippled industrial nations.

It might not happen. But the possibility exists, while the danger increases. Control of the world's most important source of energy, backed by the enormous wealth this creates, arms the Middle East and the Islamic nations with awesome power. The political implications of this, warned former U.S. Secretary of State Henry Kissinger in 1974, were "ominous and unpredictable. Those who wield financial power would sooner or later seek to dictate the political terms of new relationships."

Other fuels must ultimately provide the bulk of man's need for energy. But there are no alternatives to oil and natural gas as the twin dominant fuels for at least the remainder of this century. During this period, there will be nothing more crucial to the secu-

---

*I distinguish the non-Communist world, not for any reasons of ideology, but because both the Soviet and Chinese blocs are self-sufficient in petroleum. Russia exports both oil and natural gas to western Europe. China, with an enormous petroleum potential and a need for funds to finance its industrialization program, could emerge as a significant oil exporter. It would be ironic if the non-Communist nations were to reduce their dependence on Middle East oil by becoming dependent on Chinese oil.

1

rity of energy supplies—to the security of the non-Communist world —than the development of oil and gas production from new supply areas. Without this the non-Communist world will be increasingly hostage to that part of the globe centred on the Persian Gulf.

The largest potential new oil and gas supply areas in North America are Mexico and the western Arctic. These are the opposite ends of the belt of oil and gas bearing sedimentary rocks that stretches across the length of the continent. The polar and the tropical ends of this belt harbour the bulk of North America's prospective new supplies of oil and natural gas.

The western Arctic has already yielded the continent's largest oil field, at Prudhoe Bay on the North Slope of Alaska. Almost as much energy is contained in the natural gas so far discovered in the western Arctic. Important oil finds have been made in the Mackenzie River Delta area and the adjacent Beaufort Sea. Only a small fraction of the potential petroleum resources of the western Arctic have thus far been found by a relative handful of exploratory wells.

The most—perhaps for much of it the only—feasible means of securing oil and gas supplies from the western Arctic is by the most direct, shortest and cheapest pipeline route, which follows the sedimentary belt across northern Alaska and Canada.

Twice this route has been rejected, for reasons of political expediency. It was rejected first for the transportation of oil from Prudhoe Bay, and later for the transportation of natural gas from Prudhoe Bay and the Mackenzie Delta. The result has been an increase in North America's dependence on Middle East oil, and a decrease in the security of energy supplies.

The rejection of this route entails a sweeping range of implications. It means that none of the natural gas in the western Arctic— one-eighth of the combined proven reserves of Canada and the United States—is now available, and plans for transporting it remain uncertain. It means that billions of dollars worth of oil in the Mackenzie Delta and adjacent Beaufort Sea are now shut in because of a lack of transportation facilities. It means that the oil which is flowing from the western Arctic—American oil from Prudhoe Bay—is flowing to the U.S. west coast where it is least needed, while an effective way to get it where it is most needed, in the U.S. midwest and northeast, has yet to be determined. It means a stagnation in the effort to find and develop new resources in North America's most promising petroleum province. Lacking a way to

2

transport and sell any new oil and gas they might find in the western Arctic, oil companies have diverted hundreds of millions of dollars in exploration expenditures to other areas, most notably Alberta. It means less new energy, since for each exploratory well drilled the amount of oil and gas found in the Western Arctic is as much as fifty times greater than in western Canada. It means a severe cut back in efforts to find and develop petroleum resources owned by the Government of Canada, in favour of petroleum resources owned by the Government of Alberta. For the people of the Northwest Territories, it means the loss of the largest available opportunity for increased employment, and instead increased reliance on a precarious hunting-trapping-fishing economy. For Canada's northern native people, it means a loss of opportunity to share in oil and gas revenues through a settlement of outstanding land claims, in contrast to more than one billion dollars now being paid to Alaskan native people. It means a loss of potential tax and royalty revenues for the Government of the Northwest Territories, which will leave it dependent on the Government of Canada for more than eighty per cent of its revenues. It means that during a ten-year period tax payers in southern Canada will have to spend some \$5 billion to support government services in the Northwest Territories.

All this results from rejecting the dictates of geography and economics in the decisions by Canada and the United States respecting the transportation of oil and natural gas from the western Arctic.

For nearly a decade I was a close observer of the ill-fated plans for pipeline transportation, first of oil and then of natural gas, from the western Arctic along the only direct route that exists. With the discovery of the Prudhoe Bay field in 1968, I watched the drama of conflicting oil transportation proposals as editor of Oilweek magazine, published in Calgary. I was the first to describe in print the advantages of the direct pipeline route along the path of the sedimentary belt. I watched as the oil and pipeline companies in Canada promoted the trans-Canada pipeline route. I saw regional U.S. interests battle each other for either a trans-Canada or a trans-Alaska pipeline and tanker route. I saw Ottawa vacillate between the political horrors of acquiescing in oil tanker shipments off the coast of British Columbia, and the difficulties in quickly approving construction of a pipeline across northern Canada. It was easy to see the short-term thinking of the oil companies, desperate for revenues from their big Prudhoe Bay reserves, as they opted for the longer and more costly transportation route, which could not connect the total supply area, rather than face the risk of delay in securing approval for a trans-Canada route.

I was more closely involved in the competing proposals to transport natural gas from the western Arctic, working for more than six years as director of public affairs for Canadian Arctic Gas Pipeline Limited and its predecessor, the Northwest Project Study Group. Arctic Gas proposed to follow basically the same rejected route as the ill-fated trans-Canada oil pipeline from the western Arctic. Its proposed route would have connected with existing and new pipelines which in turn would have made the Mackenzie Delta gas available to Canadian consumers from British Columbia to Quebec, and the Alaskan gas available to U.S. consumers from the Pacific to the Atlantic seaboards.

After a decade of studies and public hearings, and expenditures of more than a quarter of a billion dollars by industry and government, the concept of transporting both Canadian and U.S. petroleum resources from the western Arctic along a single pipeline route was once again rejected. Approved instead was a pipeline route to transport only U.S. gas from Prudhoe Bay and the Alaskan North Slope along the route of the Alaska Highway. Later construction of a second, connecting pipeline is required to transport the Canadian gas from the Mackenzie Delta and Beaufort Sea. This involves a minimum of an additional one thousand miles of pipeline as compared with the single pipeline route.

Eighteen months after the Canadian and U.S. governments approved the Alaska Highway pipeline its sponsors announced that the target date for completion had been deferred by at least twenty-two months. When, or if, it may ultimately be built, remains uncertain.

But costs had nothing to do with the final decision. Public perceptions and political pressures had everything to do with the final outcome.

In this book I focus on the debate precipitated by the Arctic Gas proposal, and the issues involved in that debate. For a period in the mid-1970s, it was the major public issue in Canada. Much of that debate was irrelevant to any determination of the best means to transport a new supply of energy. The debate provided a highly visible national platform for many impassioned and often just

4

causes: for northern native people who sought a just settlement of their outstanding land claims and a means to restore dignity and self-worth; environmentalists concerned with protecting the northern wilderness; nationalists—or economic isolationists, if you will—in both countries, demanding either "all-American" or "all-Canadian" projects; advocates of a conservor society seeking simpler and more egalitarian life styles; radical activists (some within the churches) seeking to reform society along Marxist lines. All had their say in the great pipeline debate.

If in my account of the arguments and the issues, my own perceptions and assessments are discernible, I offer no apologies. I can only claim that they are mine alone, and for them I must take the entire blame. I profess no qualifications to lend authority to any views I have arrived at here, other than whatever share of common sense I may have been granted and an attempt to apply this to the issues during a prolonged period. I do not seek, nor expect, to persuade those who, with equal or greater validity, have differing perceptions.

Despite whatever tangents may have derailed it—arguments about colonialism from Asia to Aklavik, as an example—the debate was at least nominally concerned with natural gas from the western Arctic. The outcome, however, cannot be assessed without considering oil, any more than an exploration company can separate its search for natural gas from its search for oil. And no consideration of oil can ignore the Middle East, that tiny glob on the globe that accounts for seventy per cent of the discovered remaining oil reserves of the non-Communist world.

As this is written, oil exports from Iran have been cut off for some four months. Prior to November, 1978, Iran had accounted for one-fifth of Middle East oil production, roughly ten percent of total oil supplies of the non-Communist world. During a twenty-eight year period there have been at least eight partial but significant disruptions to the flow of Middle East oil, an average of one every three and a half years.

Not that it mattered much during most of this period. The world was swimming in a sea of low-cost oil. If the flow was cut from one area, the valves were simply opened a little more in other areas. When temporary shortages did occur, the invariable effect was to step up the efforts to find more oil in other regions, ultimately leading to another glut.

Even the 1973 Arab oil embargo, the watershed that marked the turning point from an era of low-cost to an era of high-cost energy, appeared to follow the pattern of shortage and surplus. There were wide shortages of gasoline and fuel oil, prices leaped with the speed an electronic computer, and the world economy slid into a recession from which it had not fully recovered more than five years later. Before long, however, Arab oil was once again flowing, stronger than ever. Important new supplies were being developed in the North Sea, Alaska, Mexico and elsewhere. The growth in oil demand was curbed by high prices, conservation measures, and the world economic recession. Service stations that closed in 1974 when they ran out of gasoline were soon once more engaged in price wars, albeit at much higher prices. As early as the beginning of 1975, writes Anthony Sampson in *The Seven Sisters*, "The signs of new plenty were evident everywhere. The storage tanks were overflowing with oil, the loading docks were half-empty, the freight rate for tankers dropped lower and lower."

While the pattern of shortage and surplus appeared to prevail, there was an important difference. The so-called surplus was now much smaller than that which had followed other shortages. Iran illustrates the point. In 1950 it supplied some twelve per cent of the oil used by the non-Communist nations, a slightly greater proportion than in 1978. When the dramatic Premier of Iran, Dr. Mossadeq, nationalized the British-owned Anglo Iranian Oil Company in 1951, Iran's oil production came to a virtual halt. It was seven years before production was restored to the 1950 level. The rest of the world hardly noticed. There was plenty of oil. But with the cut of Iran's oil production in late 1978, it has taken less than four months to feel the pinch. Before the end of February, oil companies had started to allocate fuel oil supplies, and U.S. Energy Secretary James Schlesinger was publicly warning of gasoline shortages.

The fact is that the reported oil glut which preceded this latest cut in Iranian production was more apparent than real. Oil is by far the largest commodity in international trade, and it is impossible to keep supply and demand in perfect balance. The day that there are no surplus supplies of oil is the day that shortages start.

Whatever happens in Iran and the Middle East will have a bearing on what happens to the large potential oil and gas supplies of the western Arctic. A likely result of the worst supply scenario would be a crash program to get western Arctic oil and gas on stream as fast as possible, with scant regard for economics, the

6

environment or social considerations of anyone standing on the path of the pipelines. With the alternative scenario of full restoration of Middle East oil exports and the re-appearance of surplus oil and gas supplies, prevailing public and political attitudes to matters of energy security could once again be marked by apathy, and little heed would be paid to any apprehensions of the next crunch. But for those who listen, that ticking sound in the Middle East can clearly be heard.

Woodville, Ontario   Earle Gray
February 28, 1979

PART ONE
THE CORPORATE SECTOR

Chapter 1
Moving It

From a satellite photograph of North America, you can almost see where the great interior belt of sedimentary rocks lies.

It spreads east of the Rocky Mountains across the flat lands of the continent, from the Gulf of Mexico to the Mackenzie River Delta on the Arctic coast and eastward along the narrow coastal plain to the northwest tip of Alaska, a distance of nearly three thousand miles from New Orleans to Point Barrow.

Time and again, great seas swept across where now there are plains and forests, where cattle graze in Texas, where wheat fields wave across Saskatchewan, where stunted black spruce grow on the tundra of the Arctic atop the permanently frozen ground called permafrost. Sediments of mud, silt, sand and shale were deposited at the edges of these seas, burying tropical vegetation. Great ocean reefs, hundreds of feet thick, stretching for miles, were built from the shells of tiny coral animals, stomatoporoids. Inch by inch as the continent evolved these sediments were deposited on the granite crust of the earth, compressed into the sedimentary rocks that may be as thick as five miles or more.

The sedimentary rocks that lie in this enormous trough across the length of the continent are the location of most of the oil and natural gas in North America. More than half the energy produced in Canada and the United States is extracted from tiny pores in these rocks, providing the basic supply of fuel and heat for the most energy-intensive economies in the world. From wells in these rocks, natural gas is fed through a spider's web of pipelines, covering more miles than all the railway tracks on the continent, to millions of homes, schools, factories, petrochemical plants.

The southern half of this great sedimentary belt is also the most intensively explored oil and gas region in the world. Hundreds of thousands of exploratory wells—wildcats—have pock-marked the belt in search of oil and gas accumulations from northern Alberta to a hundred miles offshore in the waters of the Gulf of Mexico. Wildcatters will continue to search here for perhaps several more decades, but most of the oil and gas has already been found. The principal source of energy that has fueled the economies of both nations is steadily shrinking.

It is the northern half of this sedimentary belt which, to a great extent, will shape the future prospects for oil and gas supplies for both countries. Most of the northern part of the belt is still virgin territory for the wildcatters. In the fourteen hundred mile span from Point Barrow to the northern borders of Alberta and British Columbia, only a few hundred wildcats have been drilled. The results have included both frustrating disappointment and spectacular success. Prudhoe Bay, on the coastal plain of Alaska, is the largest oil and gas field in North America. Substantial reserves of natural gas, as well as several billion dollars worth of oil, have been found in the Mackenzie River Delta. Farther offshore, under the deeper waters of the Beaufort Sea, the first wildcats indicate the promise of much larger reserves of both oil and gas.

In mid 1977, the most costly transportation system ever built began operating from the Prudhoe Bay field in defiance of logistics, economics and geography. The nine billion dollar Alaska pipeline was built to move two million barrels of oil a day eight hundred miles across Alaska to a Pacific coast terminal for tanker shipment to American westcoast ports. It spurns the route of the sedimentary belt, the shortest and most direct access to most of the United States. It does not connect with the oil reserves in the Mackenzie Delta and Beaufort Sea, which remain locked in for lack of transportation. Supertankers are unable to deliver Alaskan oil to those parts of the United States that need it most. Billions more must be spent to get Alaskan oil to U.S. cities east of the Rockies.

The first oil had hardly started to flow through the new pipeline before the governments of Canada and the United States reached a decision on an even more costly means to move natural gas from the same region of the western Arctic.

It is unlikely that more time and money has ever been spent to reach a decision. It required nearly a decade. It cost more than one quarter of a billion dollars for studies by industry and government and for eight hundred days of public hearings in both countries.

12

As with Alaskan oil, the decision on how to move western Arctic gas once again flaunted the logic of geography and economics, because it will cost more to supply the gas than the gas is now worth. It took a decade to make the decision, but it may take longer to implement it—if it can be implemented at all.

In the final analysis, public and political perceptions frustrated the logic of geography which should have dictated how to move the gas from the western Arctic. Contributing to these perceptions were environmentalists with a cause celebre and radical reformers who sought the establishment of a separate nation in northern Canada, a native society based on Marxist ideology.

All the government bodies that considered the matter were agreed about the way that both Canadian gas from the Mackenzie Delta and American gas from Prudhoe Bay could be transported at the least cost. Every assessment confirmed that the most economical way would be to transport the gas in a single pipeline along the shortest route, along the path of the sedimentary belt where more oil and gas fields will be found.

The decision, finally, was to build a pipeline to transport only the Alaskan gas along a route that is three hundred miles longer, across Alaska and northern Canada in the vicinity of the Alaska Highway. The route lies across a sea of mountains, well south of the flat lands of the sedimentary belt. A second pipeline, at least seven hundred miles in length, would have to be built to connect Canadian gas fields in the Mackenzie Delta with the Prudhoe Bay pipeline. Even then, after tens of billions of dollars had been spent, there would still be no pipeline to connect oil and gas supplies along most of the northern length of the sedimentary belt.

There were two reasons for the decision to use the longer and more expensive transportation routes for this natural gas. It was felt that by following existing transportation routes there would be less risk of environmental damage. It was also felt that building a pipeline now up the Mackenzie Valley, and the opportunities for wage employment that this would provide, would disrupt the life-style of native peoples in the region and thwart their quest for an alternative society based on hunting and trapping.

But if the selected routes pose no environmental risks along most of the northern sedimentary belt, neither do they provide any means of transporting gas from most of this area. And if gas discoveries in the future are ever connected by pipeline, the environmental risks will not have been avoided. They will, instead, have been compounded.

The quest for an alternative native society in the Mackenzie Valley may well prove a futile and dangerous illusion. The goal of a separate nation based on Marxist ideology is doomed to failure so long as Canada remains a nation. But the struggle has yet to be abandoned.

What has become clear is that the longer pipeline route to transport a smaller volume of gas to American cities cannot now be economically built. Cheaper sources of energy are available, from oil, from coal, from imported liquefield natural gas. American consumers have already been offered as large a supply of gas from Mexico at half the present price of Alaskan gas.

The decision that required ten years and a quarter of a billion dollars may really be no decision at all. Canada and the United States have yet to decide on how they can economically use the enormous natural gas reserves in the western Arctic.

Chapter 2
The Prize on the North Slope of Alaska

The North Slope of Alaska sweeps down from the peaks of the Brooks Mountains, tumbles across the foothills and spreads over a table-flat plain to the edge of the Arctic Ocean. It stretches three hundred and eighty miles across the northwest corner of the continent, varying from forty to one hundred and fifty miles in width. So flat and bleak is the coastal plain that on my first visit there I could not tell when we were flying over snow-covered land or the snow-covered ice of the Arctic Ocean. The only trees in sight are the assemblies of pipes and valves atop the oil wells that oil men call "Christmas trees".

Geologists had long suspected that large accumulations of oil and gas might be found in the sedimentary rocks underlying the foothills and plains in this land of desolate grandeur. In 1923, the American Government set aside thirty seven thousand acres at the western end of the slope calling it Naval Petroleum Reserve No. 4, hoping to establish a strategic supply of oil. Some eighty exploratory wells drilled by the U.S. Navy in the 1940s and 1950s found a few small pools of oil and gas. Outside this reserve, oil companies had drilled half a dozen deep and expensive wildcats, each of them dry and abandoned.

In 1967, Atlantic Richfield Company and Exxon selected two more drill sites on leases purchased from the State of Alaska. The first of these tests, ARCO Suzie Unit No. 1, cost four and a half million dollars, and was another dry hole. The drilling rig was shifted to the second location for one more try.

"It is anticipated that little or no geophysical or geological work will be conducted on the Arctic slope during 1968 unless the current drilling well is successful," predicted the 1967 annual review issue of the *Bulletin of the American Association of Petroleum Geologists*.

By the time this had been delivered in the mail, Atlantic Richfield and Exxon had discovered the Prudhoe Bay oil and gas field.

ARCO Prudhoe Bay State No. 1 in January, 1968, had drilled to eight thousand seven hundred feet, penetrating a section of sandstone that was four hundred and seventy feet thick. Natural gas was trapped in the pores of most of this rock, but the bottom seventy feet, the company stated, was "believed oil-saturated on the basis of core examination." The drilling rig was shifted to a second drill site, ARCO Sag River, located down the flank of the structure and intended to determine if there was more oil beneath the gas in the sandstone formation. On June 25, Atlantic Richfield announced that its Sag River well had "encountered oil in the same Triassic formation," penetrating some three hundred feet of oil-saturated rock.

The first published report to suggest the possible magnitude of this discovery was written three weeks later by *Oilweek's* exploration editor Horst Heise. He speculated that if the sandstone contained oil throughout the area of the structure, "it would be one of the biggest fields on the continent and the richest Arctic find in the world" with reserves "in the two to seven-billion barrel range." This would mean that this single field could contain more than half the amount of oil in hundreds of separate fields discovered in Canada.

Several weeks later, Atlantic Richfield released an assessment prepared by DeGolyer and MacNaughton, consulting petroleum geologists and engineers. On the basis of these first two wells alone, the consultants concluded: "In our opinion this important discovery could develop into a field with recoverable reserves of some five to ten billion barrels of oil, which would rate as one of the largest petroleum accumulations known to the world today." By the time subsequent wells had fully delineated the field, recoverable reserves were estimated at a minimum of ten billion barrels of oil and nearly thirty trillion cubic feet of gas. In terms of energy supply, the gas is equivalent to more than five billion barrels of oil.

How could these reserves be transported from this remote corner of the continent to markets where it could be used?

While Horst Heise speculated in the July 22, 1968, issue of *Oilweek* on the size of the Prudhoe Bay discovery, I speculated in the same issue on the possible means of transportation. There were, I wrote, three options: by tanker through the ice-covered Arctic waters; by pipeline eight hundred miles across Alaska to a Pacific coast port and by tanker shipments from there; or by a sixteen hundred mile pipeline following the belt of sedimentary rocks east from Prudhoe Bay along the coastal plain to the Mackenzie River Delta and across the Northwest Territories and Alberta to Edmonton. At Edmonton, such a pipeline could connect with the most efficient oil pipelines of the continent, stretching southeast to Chicago and Toronto, and southwest to Vancouver and Seattle.

A trans-Canada route for Alaskan oil offered two major advantages, I wrote. It would provide a means to transport any oil found along the northern sedimentary belt from Prudhoe Bay to Alberta. It would also move Alaskan oil to "that area of the United States which faces the largest petroleum deficiency." By contrast, "The most serious problem facing shipment of this oil to the Pacific coast," by means of a trans-Alaska pipeline and tankers, "is that this is the market region where the oil is least needed."

The idea of a pipeline to move Alaskan oil across Canada attracted support from a number of quarters in Canada. Exploration companies saw it as the key to transporting the oil they hoped to find in the north. Ottawa saw it as a means of developing federal oil and gas resources and averting the risk of pollution by oil spills from tankers moving Alaskan oil along the British Columbia coast and through the narrow Juan de Fuca Straits to Puget Sound refineries in the state of Washington. Before the end of 1968, Trans Mountain Oil Pipeline Company had completed a preliminary study of the trans-Canada route and concluded that an oil pipeline looked feasible. By 1971, Canada's Energy Minister Joe Greene was publicly advocating that oil companies scrap their plans for a trans-Alaska line in favour of a trans-Canada line which, he said, would deliver the oil at less cost, follow an ecologically preferable route, and avoid the risks of tanker spills.

The three companies with most of the oil at Prudhoe Bay (Atlantic Richfield, Exxon and British Petroleum) had set their sights on a pipeline across Alaska, however. The main attraction was that a trans-Alaska pipeline could be put into operation much sooner, in part because it involved only half as much construction, and also because it seemed to avoid the risk of lengthy delays in negotiating a pipeline route across Canada.

In February, 1969, the three Prudhoe Bay producers issued a joint statement saying that they had decided to proceed with a trans-Alaska pipeline, while continuing studies of a second possible transportation route. The trans-Alaska line, they figured, would cost about nine hundred million dollars and could be completed by 1972. Some two thousand men would be involved in construction. Forty-eight inch diameter pipe had already been ordered from Japanese steel mills and started arriving in Alaska later that year. "While the planned line to the Gulf of Alaska would make north slope oil available to the U.S. west coast, planning studies on alternative routes are continuing to determine the best method to move the oil to the midwest and eastern sections of the United States," the companies stated in a joint announcement.

The trans-Canada, Mackenzie Valley pipeline route was examined more intensively during a three-year period by five companies (Trans Mountain, Atlantic Richfield, Imperial Oil, Shell Canada and Gulf Oil Canada) who organized Mackenzie Valley Pipe Line Research Ltd. After the decision to route Prudhoe Bay oil across Alaska was confirmed, further studies were made of a pipeline which would move oil from just the Mackenzie Delta, where the first oil discovery had been made by Imperial Oil in 1971. These studies were later suspended, pending the discovery of enough oil to economically justify such a pipeline. Had Alaskan oil been routed across Canada, production from the Canadian reserves could have started in 1977. As it was, by 1978 there was more than six billion dollars worth of discovered oil reserves locked in the Mackenzie Delta, with indications of much more farther offshore under the Beaufort Sea, and no means to transport it.

Studies of the other alternative for moving Prudhoe Bay oil, shipping it out by tanker, involved the voyages of the SS Manhattan through the Northwest Passage in 1969 and 1970. Humble Oil (then the name of Exxon's U.S. operating company) had visions of thirty supertankers each carrying two hundred and fifty thousand tons of oil plying the Northwest Passage from Prudhoe Bay to the U.S. east coast by 1980. "An open Northwest Passage means not merely an oil route, but an international trade route that will have profound influence on the rate of Arctic development and the pattern of world trade," declared Humble's president Dr. Charles Jones. "A year-round sea route in this area could do what the railroads did for the western United States—and might do it quicker."

Aside from Texas hyperbole, the thought of doing to the Arctic everything that the railroads did to the western United States was not greeted with universal acclaim. But Humble intended to find out if its grandiose ideas were feasible, and to do so it took a two year charter on the largest, most powerful tanker in the U.S. fleet. The SS Manhattan was cut into four sections, and the parts were towed to four different shipyards for modifications. A new one hundred and twenty-five foot bow was designed and built and ice fenders made from high strength steel an inch and a quarter thick were added to the hull. A total of nine thousand tons of steel was added to the Manhattan, bringing her displacement to one hundred and fifty-one thousand tons—a thousand foot dreadnaught driven by forty-three thousand horsepower. Then, Texas sailors took her north to see what she could do in the Arctic.

"We believe the Manhattan has the power and the design that will enable her to bull her way through most ice ridges like an immense ram rod," predicted Humble's task force commander Stanley B. Haas.

On the two thousand mile voyage from Pennsylvania to Prudhoe Bay, the Manhattan got stuck in the ice a dozen times, freed by the relatively tiny (nine thousand tons) Canadian Coastguard icebreaker John A MacDonald. With the engines rocking his big ship back and forth to free it from the ice that gripped it like a vice, Manhattan's Captain Roger Steward drawled, "You just gotta keep 'a-swinging'."

In McClure Strait, north of Banks Island, the Manhattan was not doing much swinging. Caught in a pan of polar ice ten feet thick, with ridges twice as thick—sparkling crystal blue ice, centuries old and as strong as concrete—the Manhattan was stuck for thirty-four hours. From where I watched it, as a journalist back and forth between the two vessels on this journey, it seemed to me that the Manhattan would be there yet if it were not for the help of the Johnny Mac. Still, the Manhattan did demonstrate that, if properly designed, an icebreaker tanker could operate through these Arctic waters.

"The 4,500-mile voyage of the supertanker SS Manhattan has established that cargo shipping through the Northwest Passage is operationally feasible," I wrote in *Oilweek*. "The question now is whether or not it will be economically feasible."

A year later, after a second voyage into polar ice and expenditures of fifty million dollars, Humble announced its verdict: "The use of icebreaker tankers to transport crude oil from Alaska's north slope to U.S. markets is commercially feasible . . . but pipeline transportation appears to have an economic edge." Putting the best possible appearance on the matter, the company statement concluded: "The two Arctic voyages of the SS Manhattan were highly successful in providing valuable data for our studies concerning the various transportation alternatives for moving Alaskan crude oil to U.S. refineries."

While Humble set out on its Manhattan voyage, things were humming in Alaska. In a rush to delineate the size of the Prudhoe Bay oil field, in the spring of 1969 the companies were flying drilling rigs, cement, mud, timber, fuel and other supplies from Fairbanks four hundred miles north to the slope in the largest commercial airlift ever undertaken. Eleven giant Lockheed Hercules airfreighters (five of them brought in from Zambia where they had been flying copper to Tanzania) were flying around the clock, each carrying up to thirty thousand pounds of freight. In March alone, more than a thousand round-trip flights were made.

On September 10th of that year, scores of oil men crowded into the Sydney Laurence Auditorium in downtown Anchorage, nervously holding their brief cases which contained sealed bids and certified cheques totalling billions of dollars. They had come to bid on one hundred and seventy-nine leases covering four hundred and fifty thousand acres on the North Slope which the state had put up for auction. By the end of the day, the State of Alaska had collected nine hundred million dollars, at that time the largest amount received in a single sale of oil and gas rights.

Bouyed by this windfall and the prospects of further large revenues from its twenty per cent royalty on North Slope oil production, the Alaska State Legislature approved a $314 million budget for fiscal year 1970-71. That was more than double the state's budget for the previous year.

Then it started to dawn on the legislators that the royalty cheques might not be coming in as soon as they had hoped. Five groups: the Friends of the Earth, the Wilderness Society, the Environmental Defence Fund, the Cordova District Fisheries Union, and the Canadian Wildlife Federation had applied for an injunction under the new U.S. National Environmental Protection Act to halt planned construction of the pipeline. Judge George Hart, Jr. of the U.S. Federal District Court in Washington granted the injunction in April 1970. At the same time, Interior Secretary Stewart Udall slapped on a land freeze, suspending mineral leasing, land transfers and the granting of rights-of-way and special use permits on public

20

lands in Alaska, pending the settlement of the land claims of Alaska's native people.

Concerned by this turn of events, in April the State of Alaska chartered a DC8 to fly one hundred and twenty Alaskans to Washington. There they lobbied key congressmen and Interior Department officials to try to get the pipeline approved, but to little avail.

From there, things seemed to get worse and worse, as the pipeline got delayed and delayed, while the cost went up and up. By 1970, the cost had risen from the initial guess of nine hundred million to one and a half billion dollars. The estimate rose to three billion in 1972, four billion by 1974, six billion by 1975, and still climbing.

The land freeze was lifted in 1971, after Congress passed the Native Land Claims Act, which gave Alaska's sixty thousand native people title to forty million acres in the state, plus one billion dollars to be paid over a period of time, in part from royalties on North Slope oil production. The Alaska Federation of Natives quickly joined other state interests in lobbying for approval and construction of the pipeline.

By March of 1972, the U.S. Department of the Interior had spent ten million dollars studying the environmental implications of the pipeline and had filed a nine volume Environmental Impact Statement. Judge Hart studied the department's report for five months, agreed that the pipeline could be built without intolerable environmental effects, and lifted the injunction that August.

The environmentalists were not through yet, however. A stipulation of a 1920 federal Mineral Leasing Act provided that no pipeline right-of-way across public lands could exceed twenty-five feet in width. With the big equipment required to construct the Alyeska pipeline, a much wider right-of-way than that was required, and Interior Secretary Rogers C. B. Morton had promised a special permit for a one hundred and fifty-foot right-of-way. The U.S. Circuit Court of Appeal in February 1973 enjoined Morton from issuing this special permit, ruling that Congress would have to first either amend the law of exempt this project from the law's provisions. Once more the Alyeska pipeline was stalled, its fate again in the hands of Congress.

This latest delay in approval of the Alyeska pipeline provided a last-gasp chance to promote the alternative pipeline route across Canada.

42683

David Anderson, first as a Liberal Member of Parliament from British Columbia, later as the leader of the provincial Liberal Party, crusaded vigorously for the trans-Canada pipeline route. By March 1973, he had lined up support in Washington from six Senators and thirty-two Representatives who called on President Nixon to open negotiations with Canada for a trans-Canada oil line. "We should be as helpful as we possibly can to the Americans," Anderson said. "We should do everything we can to provide them with their oil." Senator Walter Mondale, later Vice President Mondale, introduced a bill to approve widening of the pipeline-right-of-way, on the condition that it follow a route across Canada. Mackenzie Valley Pipe Line Research Ltd. presented the results of its economic and feasibility studies to the federal cabinet in Ottawa early that year. The Independent Petroleum Association of Canada urged support of the trans-Canada route, claiming that it "would assist materially in development of the region and make marginal oil finds in the Canadian north an economic possibility."

There is little doubt that Ottawa was not happy about the prospects of oil tanker shipments off the British Columbia coast and would have preferred a Mackenzie Valley route. The trouble was that Ottawa was not in a position to assure Washington that such a route could be made available, and certainly not without substantial delay. No public hearings had been conducted on the environmental implications of such a route, and Ottawa was far from having settled the land claims of native people in northern Canada. To have approved a Mackenzie Valley oil pipeline without first addressing these two matters would have been politically intolerable. By failing to focus on these issues much sooner, Ottawa lost the opportunity to even determine whether such a pipeline would have been in Canada's national interest.

It was the Arab oil embargo in the fall of 1973 that finally jolted Congress into passing a right-of-way bill. President Nixon signed the bill in November, and a crash construction program was begun. By the time the line was completed, five years behind schedule, the cost amounted to nine and a third billion dollars, including the interest charges.

These sky-rocketing costs came close to bankrupting the oil company founded by John D. Rockefeller. The Standard Oil Company, of Cleveland, Ohio (now more commonly known simply as Sohio) was founded by Rockefeller in 1870, the first of a string of Standard Oil companies which he later controlled. When the United

States Supreme Court split control of the Standard companies under anti-trust legislation in 1911, Sohio was one of the smaller companies to emerge, a regional refining and marketing operation with a limited supply of oil. Fifty years later, Sohio still had to buy three quarters of the oil needed for its refineries from other companies, a situation it had long and unsuccessfully sought to remedy.

Sohio's opportunity to acquire a significant oil supply came with the Prudhoe Bay discovery. British Petroleum, with leases near the Atlantic Richfield discovery well, wound up with more than half the oil reserves in the field. In 1969, Sohio negotiated to purchase BP's U.S. interests, including the Prudhoe Bay oil, in return for Sohio shares. The deal gave the English company more than half ownership in Sohio. (BP, in turn, is owned fifty-one per cent by the Government of Great Britain.)

Sohio's problem was then to borrow enough money to pay for its share of the cost of the pipeline and other heavy expenditures that it faced. As the pipeline kept getting delayed and the cost kept going up, the problem got tougher. Sohio was soon faced with two alternatives. It could sell the oil reserves that it had purchased at Prudhoe Bay and, noted *Fortune* magazine, other companies "were waiting like wolves at the door, eager to pick up a piece of Sohio's position." Or it could borrow every penny it could. It decided to borrow. The risks in this were whether it would be able to borrow enough and whether it could afford to carry the debt until the pipeline came on stream. By the time the pipeline was completed, Sohio had borrowed $4.6 billion, fifty times its debt in the late sixties. At one point, when the pipeline was still being delayed, Sohio was down to where it had only enough money to meet its payroll and other expenses for six weeks, unsure whether it would be able to borrow any more. One of the institutions which came to Sohio's aid, says *Fortune* (August, 1977) was, "the swinging Bank of Nova Scotia," which provided a hundred million dollar revolving line of credit and gave Sohio some breathing room.

When the pipeline was completed, Sohio emerged as the third largest oil producer in the United States. But the only thing that had saved it from financial disaster was the sudden burst in world oil prices following the 1973 Arab oil embargo. Even then, when the line was completed Sohio still faced the task of paying up to $600 million a year in interest and repayments on the debts it had accumulated.

It required nine years from the time of discovery to build the transportation system and put Prudhoe Bay oil reserves into production. It took the same nine years merely to decide how to transport the natural gas. The problems and financial risks involved in constructing a gas pipeline will be even more severe.

Chapter 3
A Vision of the Main Gas Artery

Competition between gas pipeline companies is often for some of the biggest public franchises issued to private investors. As a public utility, a gas pipeline owes its very existence to winning a government franchise, its every aspect of operation controlled by government regulatory bodies.

Political competition can loom almost as large as market competition. Almost. Because, even with a franchise, a pipeline that cannot meet market competition from other fuels is headed for financial disaster. In the past, this element of business risk has seldom been great for gas pipeline companies.

In the United States, natural gas came of age immediately after the Second World War. For millions of Americans it became nature's new wonder fuel, clean, efficient, convenient, dependable. No more chopping wood, shovelling coal, carrying sawdust or waiting in the cold for a delivery of fuel oil. Industry, too, turned rapidly to gas. It was cheap, the supply seemed limitless, the demand insatiable. Fierce battles were fought in front of regulatory agencies, and the prize was the right to build the big pipelines. It was the growth industry of the era, and natural gas was well on its way to becoming the second major source of energy used in the United States, at its peak supplying one-third of the total.

Natural gas supplied only two per cent of all the energy used in Canada by 1950, and nearly all of that was in Alberta. During the next two decades, the American pattern was repeated. Canada was girded with pipelines carrying western gas from Vancouver to Montreal, and to U.S. markets as well. By the early 1970s, natural gas supplied one-fifth of all the energy used in Canada.

The picture had changed by the time large gas reserves were discovered at Prudhoe Bay and, later, in the Mackenzie Delta. In the United States, the insatiable demand had caught up to the limitless supply. Under government regulation, gas was still being sold cheap, too cheap. The gas supply was being burned up at a faster rate than new reserves were being discovered. Industrial customers were rationed. Potential new gas customers in many cities were refused service. Canada, starting to worry about its own supplies from Alberta and British Columbia, had called a halt to further increases in its gas exports to the United States.

For the gas pipeline industry, the large new reserves in the western Arctic appeared to mean several things. They seemed to offer important help in alleviating growing U.S. shortages and an enormous investment opportunity. A pipeline to carry this gas would have to be big. It would become, predicted Sidney Robert Blair, president of The Alberta Gas Trunk Line Company, the principal gas pipeline system in North America. It would be the largest public franchise ever awarded to private investors. It would also be the largest risk ever assumed by private investors. No longer would it be safe to ignore market competition in the heat of political competition.

THE COMPETITORS

The competitors for this biggest franchise of all aligned in groups that changed with the ebb and flow of the battle.

Leading the challenge to the giants of the gas pipeline industry, as well as major oil companies, was The Alberta Gas Trunk Line Company. Originally formed as a provincial utility, prohibited by its charter from operating outside of Alberta, Trunk Line was inexorably launched on a course of conflict with most ot the industry.

Other leading corporate players in the drama included Canada's two major gas pipelines, TransCanada PipeLines and Westcoast Transmission. Westcoast and TransCanada were among five groups that had applied in 1950 for the first permit to move natural gas from Alberta. TransCanada sought approval for the world's longest pipeline, two thousand two hundred miles from Alberta to Montreal. Westcoast sought a shorter, six hundred and fifty mile, pipeline from northern Alberta and British Columbia to supply gas to BC centres and the U.S. Pacific northwest.

Westcoast won the first permit in 1952. But after two years of hearings in Washington before the Federal Power Commission, Westcoast lost out on the bid for a franchise to supply gas to the U.S. Pacific northwest. Instead, the FPC awarded approval to a fourteen hundred mile pipeline from New Mexico proposed by Pacific Northwest Pipeline Corporation.

Without U.S. sales, the Westcoast line was not economical. The FPC decision looked like the end for Westcoast to everyone but Francis Murray Patrick McMahon, the two-fisted diamond driller, promoter, gambler and wildcatter from Moyie, BC. McMahon had played for broke before. His biggest gamble had been in 1930 when he put down his last hundred dollars on an option to acquire a $20,000 lease near the Turner Valley field and then set out to find the remaining nineteen thousand nine hundred, plus enough to drill a well. That had been the start of Pacific Petroleums Ltd. which, with a string of further McMahon gambles, had grown to become one of the largest independent oil exploration companies in Canada, and sponsor of the Westcoast project.

McMahon refused to accept defeat. Pacific Northwest Pipeline had the franchise, but it did not have enough gas supply. McMahon had the gas supply, but no franchise. By offering to sell gas to Pacific Northwest at a low price—lower than the price charged to Canadian customers—McMahon was able to make a deal. Both pipelines were built. Completed in 1957, the Westcoast line was Canada's first major gas transmission system.

McMahon and his associates made fortunes. They purchased six hundred and fifty thousand Westcoast treasury shares at five cents each. Two years later, after government approvals had been secured, Westcoast shares were offered to an eager public at five dollars each. But by 1960 forty-five per cent ownership of Pacific Petroleums, and thus effective control, had been acquired by Phillips Petroleum Company of Bartlesville, Oklahoma. Through Pacific, Phillips also controlled the largest block of Westcoast shares. In 1978, control of Pacific Petroleums was purchased by Canada's state-owned oil company, Petro-Canada.

By the late 1960s, the Westcoast system had been extended farther north to tap the first gas supply brought out of the Northwest Territories, from the Pointed Mountain field near the junction of the BC, Yukon and NWT boundaries. By 1969, Westcoast was starting to look toward the biggest undertaking of them all, a gas pipeline from the North Slope of Alaska.

27

McMahon's former competitor and now Westcoast's largest customer, Pacific Northwest Pipeline, meanwhile underwent a metamorphosis, to finally emerge as a partner with Westcoast and Trunk Line in the Alaska Highway pipeline project. Three years after it started operations, Pacific Northwest Pipeline was purchased by El Paso Natural Gas Co., the largest U.S. gas utility. The deal precipitated one of the longest anti-trust and divestiture proceedings on record. The American Justice Department moved to block the El Paso merger on the grounds that El Paso and Northwest were potential competitors. The American Supreme Court ruled that El Paso had to sell its newly acquired subsidiary. It took seventeen years of protracted court proceedings and false starts before a satisfactory sale was finally approved. The winner was a group of small oil companies led by Texas oil man John G. McMillian and his Tipperary Corporation. McMillian spent more than three and a half million dollars to intervene in the suit and bid for the pipeline. By 1974, McMillian's group had acquired a twenty per cent interest in Pacific Northwest for $20 million, with the other eighty per cent held in a trust voted by Northwest's board for shareholders who held El Paso stock at the time of the divestiture. McMillian's group formed Northwest Energy Co., headquartered in Salt Lake City, Utah, as the holding company which controls the pipeline subsidiary, now named Northwest Pipeline Corporation.

Cast from the same mold as Frank McMahon, McMillian spent twenty-five years in the oil and gas business as an independent wildcatter; he likes to run his own show. "You have to have an iron fist," he is quoted as stating. "Around here, it's my way or the highway."

There was one other matter that El Paso decided to clear up when it finally completed disposing of its interest in Pacific Northwest Pipeline. It decided at the same time to sell the interest it had acquired in Westcoast Transmission. In January, 1974, the Government of British Columbia paid El Paso nearly thirty million dollars for one million one hundred and fifty-seven thousand shares of Westcoast. This purchase made the BC government the second largest owner of Westcoast with thirteen and a half per cent interest, compared with the thirty-five per cent controlled by Pacific Petroleums.

More momentous than the contest to ship Alberta gas west was the contest to ship it east. Competing for the franchise were Western Pipe Lines Limited and TransCanada. Western was backed by

Osler, Hammond and Nanton, a pioneer investment house in Winnipeg which had helped finance early railway construction and had gained title to oil and gas rights on three million acres of railway land grants in Alberta. Western proposed a pipeline as far east as Winnipeg, then south to the United States border at Emerson, where it had a contract to sell gas to Northern Natural Gas Company of Omaha, Nebraska. Lionel Baxter, president of Osler, Hammond and Nanton, told the Alberta Oil and Gas Conservation Board that he had "looked into the question of taking gas to Ontario and I could not be convinced that it was economically feasible to take a line across a thousand miles of rock and muskeg and make it pay at the other end." TransCanada, backed initially by Texas wildcatter Clint Murchison, proposed to do exactly that. The two thousand two hundred mile TransCanada pipeline would be "an all Canadian project," a House of Commons committee was told. "Canadian gas [would be] transported over an all-Canadian route," while "one hundred per cent of the consumption would be in Canadian cities."

Alberta rejected both the Western and TransCanada applications in 1952, ruling that there was not enough gas surplus to the province's own needs. Both re-applied, and both were again rejected the following year. This time the Alberta board ruled that the price at which Western proposed to sell its export gas to Northern Natural was too low, while the TransCanada line was not economical without export sales to help make it pay at the other end.

With nudging from Alberta Premier E. C. Manning and Federal Trade and Commerce Minister C. D. Howe, the two firms agreed to join forces in a continuing company called TransCanada. Within five months of the merger, TransCanada had all the permits it required from both Edmonton and Ottawa. The only thing it lacked was money. One condition of TransCanada's approval was that financing was to be secured by the end of 1954. It was a full two years later than that, after the deadline had been extended five times, before TransCanada was able to raise the money for its $375 million pipeline. The money was secured after Howe had rammed legislation through Parliament to provide an $80 million loan to TransCanada and establish a Crown corporation, Northern Ontario Pipeline Company. The Crown corporation financed construction of the $180 million section of the system across northern Ontario which was leased to TransCanada.

In ramming the bill through in a hurry, the government employed unprecedented measures to cut debate short, bending Parliamentary rules in marathon sittings that extended through the nights to as late as four forty in the morning. The pipeline debate was the stormiest in Parliament's history. In the national elections a few months later, the infamous pipeline debate was a major factor in the defeat of the Liberal government after twenty-two years in office.

The pipeline was completed in 1958, and in 1963 TransCanada purchased the northern Ontario section from Ottawa, which earned good interest on its investment. Within a few years, TransCanada was more than ninety per cent Canadian-owned. And eleven years after completing what was then the longest pipeline in the world, TransCanada was leading the race to build a gas pipeline from Alaska.

Back in Omaha, Nebraska, Northern Natural Gas Company was still seeking its first supplies of natural gas from Canada, a guest which it had pursued without success for nearly thirty years. The first effort was in 1949 when Northern had contracted to buy gas from the ill-fated Western Pipe Lines. Later attempts met with no greater success. In 1969, Northern announced plans for a $104 billion, seventeen hundred mile pipeline from the Pointed Mountain area in the southern Northwest Territories to the U.S. midwest. That fell through when the reserves at Pointed Mountain turned out to be much smaller than had been anticipated.

Beginning in 1967, TransCanada PipeLines, joined by American Natural Gas of Detroit and People's Gas of Chicago, had also been studying the feasibility of a pipeline from Pointed Mountain, and the TransCanada studies were ultimately to evolve into what became the Arctic Gas proposal.

Closely following the disappointment felt over the limited size of the reserves at Pointed Mountain was the excitement generated by the discovery of truly large reserves at Prudhoe Bay. The focal point of interest leaped a thousand miles farther northwest.

It was at this time (December, 1969) that a new actor appeared in the drama, Sidney Robert (Bob) Blair, who joined The Alberta Gas Trunk Line Company as executive vice-president, and seven months later was appointed president and chief executive officer. From the moment he joined Alberta Gas Trunk Line, Blair was determined that he would stake out a major role for the company in any pipeline built to transport gas from the western Arctic. The first task was to change the very nature of Trunk Line, a staid, conservative utility which performed a limited function in an unobtrusive but profitable manner.

There was only one purpose in mind when Trunk Line was established under the auspices of the Alberta government in 1954. That purpose was to prevent the Federal Government from exercising legislative and regulatory control over the production, marketing and pricing of Alberta gas.

It was clear that the major gas pipelines planned in 1954 to ship Alberta gas to markets from Vancouver to Montreal would be subject to the exclusive legislative control of the Government of Canada. The provincial fear was that this Federal Government control would enter Alberta with the pipelines, giving Ottawa the say in the pricing and marketing of Alberta gas, and threatening the province's ability to ration its supplies to other Canadian markets.

The British North America Act gives Ottawa exclusive authority over "works and undertakings connecting the Province with any other or others of the Provinces, or extending beyond the Provinces." To prevent this from occuring in Alberta, Trunk Line was conceived as a provincial pipeline grid which would transport the province's export gas to connections at the border with the federally regulated pipelines. The idea was to head off the feds at the border.

Calgary lawyer John Ballem, in his article "The Constitutional Validity of Oil and Gas Legislation" (*Canadian Bar Review*, 1963), discussed the unique role of Trunk Line. He noted that its Act of Incorporation "does not contain any provision to the effect that all gas destined for export from Alberta must be transported through this system, nor does any other relevant Act contain this restriction. It is a fact of economic life in Alberta, however, that no permit for export of gas from the province will be granted except on condition that it be transported within Alberta solely through the facilities of Alberta Gas Trunk Line." It is still an unresolved question, however, whether legislative control of Trunk Line properly belongs to the Government of Alberta or the Government of Canada.

When the Alberta government set up Trunk Line in 1954, the initial public sale of shares was offered only to residents of Alberta. Banks, the provincial government's treasury branches and brokers' offices were swamped with applications from nearly one hundred thousand Albertans who purchased two and a half million shares at a price of $5.25 per share. Within months the shares were selling at double the issued price. Most Albertans sold their shares and took their profits, so that the bulk of the ownership was soon held outside the province.

Under the unique share structure of Trunk Line, however, it makes little difference where the ownership resides, because ownership in this case is not control. Ownership of the company by 1977 was represented by more than twenty million issued common shares, but less than two thousand of these shares carry any voting rights. The voting shares, and thus control, are held by nominees of the Alberta government and the major sectors of the province's gas industry. It is difficult to conceive of any other corporation of similar size where the owners have so little opportunity for representation in the affairs of the company.

It is clear from the act incorporating Trunk Line that the original intent was that its function be confined to handling natural gas within Alberta. The act stated that, "The objects and powers of the company do not authorize and shall not be interpreted to authorize the purchase, acquisition, construction, operation or control by the company of any works or undertakings situate outside of the Province of Alberta."

Eight years after Bob Blair walked into the office, Trunk Line had a twenty per cent interest in a nineteen hundred mile pipeline moving gas liquids across two provinces and five states from Edmonton to Sarnia; an interest in a Saskatchewan steel and pipe manufacturing complex; a subsidiary pipeline valve manufacturing company with plants in the United States and Italy; the lead role in a billion dollar petrochemical complex under construction in Alberta; plans to ship liquefied natural gas by tanker from the Arctic Islands to the Atlantic provinces; plans for an extension of the Trans-Canada pipeline from Montreal to New Brunswick; controlling interest in Husky Oil, with production, refineries and service stations in both Canada and the United States; and the leading role in the planned multi-billion dollar gas pipeline from the North Slope of Alaska.

TRUNK LINE STAKES A ROLE

The circumstances that compelled Trunk Line to challenge much of the industry were already in place when Bob Blair joined the firm in 1969. Trunk Line's whole corporate being rested on its two

32

thousand seven hundred miles of pipeline built at a cost of $320 million to move Alberta gas to various provincial boundary points and the supply of gas to keep those pipelines operating. At some unknown point in time the supply of Alberta gas would start to diminish and Trunk Line faced the long-term prospect of gradually being phased right out of business. The emerging prospect of transporting large volumes of natural gas from the western Arctic offered an obvious solution. Trunk Line could best assure major corporate expansion, as well as extended use of its existing pipelines, if it were to control the movement of northern gas across Alberta. It was a logical and compelling corporate motivation.

Two proposals to transport Alaskan North Slope gas had already been advanced by late 1969. Both planned to by-pass both Alberta and Trunk Line. Mountain Pacific Pipeline Ltd., sponsored by Westcoast Transmission and Canadian Bechtel, proposed a route across the Arctic coast from Alaska, up the Mackenzie Valley and through British Columbia to supply Alaskan gas to U.S. west coast markets. TransCanada PipeLines and its two U.S. partners, American Natural Gas and People's Gas Light and Coke, proposed a route similar at the northern end, but then cutting across the northwest corner of Alberta and diagonally across Saskatchewan and Manitoba to the U.S. border at Emerson, south of Winnipeg. This route to the U.S. midwest all but by-passed Alberta, and completely by-passed Trunk Line.

Five years of hard-nosed competition and bargaining produced a compromise that resolved much of the conflict and temporarily brought nearly all of the interests together in the Arctic Gas project. A route to ship the gas to markets east and west of the Rockies involved a compromise of east-west interests, not only in the United States but also in Canada, since by this time Canadian gas in the Mackenzie Delta was also involved. The compromise met a basic objective of Trunk Line and Alberta by routing the line across the length of that province and provided a conditional commitment to use future spare capacity in Trunk Line's existing system. What it did not satisfy was Trunk Line's interest in corporate ownership and control, which it saw as the only means to assure that its interests would be maximized. This was the crucial point of conflict.

Although it was seldom a feature of public discussion, statements by both Blair and Arctic Gas president Vern Horte confirm that this was, in fact, the central issue of corporate disagreement. "There was never the slightest doubt in my mind that, right from the start, what Bob Blair was after was control by Trunk Line of any facilities moving northern gas across Alberta, and he remained adamantly opposed to any arrangement which did not accomplish that," Horte later stated in an interview.

Blair, as early as 1970, wrote to Alberta Premier Harry Strom about the need for "obtaining and protecting a satisfactory position" for Trunk Line and concerning the advantages of a northern gas pipeline "being conceived and run from within Alberta." Seven years later he told the National Energy Board that corporate control of the facilities in Alberta to move northern gas "was part of the whole contrast" between his approach and that of Arctic Gas.

Blair expanded on his view of this issue in an appearance before the National Energy Board on May 11, 1977, the second to last day of the board's hearings on northern gas pipelines. Under questioning by board counsel Hyman Soloway, Blair said that in the movement of northern gas it is important that "the best possible use be made" of Trunk Line and Westcoast facilities in Alberta and British Columbia. He explained that the "people in the end who can determine that the best and most economical use is made for all concerned" are Trunk Line and Westcoast. Then, in one terribly long sentence, Blair explained what had been worrying him for the past seven years: "In Alberta we have been preoccupied since 1970 with the realization that if a complete express system to move all future Canadian as well as American gas [from the north] were installed across the province it could very well pre-empt on the existing transmission company an ability to compete in moving increments of supply forever, and leave the existing company with a system whose supply would, at sometime out in the future, begin to decline, and the unit cost would start to rise." This could best be averted if Trunk Line had the corporate or management control of the Alberta section of the international pipeline carrying northern gas. According to Blair, "We thought it important that Alberta Gas Trunk Line maintain the management and control of reviewing how new increments of supply should be moved as between additions to new express systems, and as between utilization of capacity in existing systems."

Blair also stated that he saw the risks of "a real conflict between the intentions of a management which was completely express line oriented, which had no particular affinity to one local area, and the service in that local area to what was most economic. That was part of the whole contrast that eventually grew and emerged out of the Gas Arctic–Northwest Project Study group." He claimed that everybody's interest would be better served if both the existing Trunk Line system and the Alberta section of the new line were controlled by the same management.

Geoff Edge, one of the three members of the NEB's hearing panel, asked if "that policy automatically optimizes the Alberta position but does not necessarily optimize the Canadian position?"

Blair: "No, I do not think so. I think it optimizes very much the Canadian position."

Edge: "Why do you have to have provincial control in a federal pipeline to optimize the Canadian position?"

Blair: "I do not believe you should have provincial control of a federal pipeline."

Edge: "I thought you said you wanted an inter-provincial pipeline essentially controlled only within the province to ensure you optimize that situation in relation to Alberta Gas Trunk Line which now exists."

At this point, Blair and Edge were not talking about quite the same thing. What Blair had conceded (reluctantly) was that it was not necessary to have Alberta government regulatory control. He still maintained that it was necessary to have Alberta corporate and management control. He said that those responsible for the operation of the federal pipeline and the Alberta pipeline "should be the same people. I think they can perfectly wear two hats and exercise their responsibility in respect of each jurisdiction. They can be supervised by the National Energy Board in respect of the federal responsibility and by the provincial authorities in respect to the provincial responsibility." Blair claimed that "if you separate the management you then produce two possibly divergent views on what might be the best Canadian interest ... I think if they are the same people, then they do not have that problem."

What this really means is that if there is a conflict between the Canadian interest and the Alberta interest, the best people to re-solve that conflict would be Alberta Gas Trunk Line.

Soloway arched a bushy eyebrow and registered some concern about this view. Could it be that Trunk Line would resolve a conflict in favour of Alberta interest at the expense of Canadian interest? He suggested to Blair that "the conflict you are describing might work the other way, rather than the way you anticipate it would work."

Blair responded: "What I am speaking of doing is making sure the interest which management has represents the interest of doing an economical job in Alberta as well as doing an economical job of this international transmission service."

Soloway: "Does an economical job in Alberta involve the utilization of facilities of Alberta Gas Trunk which may become surplus at a later date?"

Blair: "It could very well."

"If certain problems did arise, would they not be matters for the National Energy Board to resolve, after hearing all points of view?" Edge asked. "Yes," said Blair, but it would be better if they were resolved by Alberta Gas Trunk Line rather than by the National Energy Board. "It is very hard to take negotiations before a regulatory board to get fair treatment," Blair said. "I would sooner see that worked out by management in the first place." And, of course, if Trunk Line controlled both the provincial and inter-provincial systems, then it could resolve these little problems itself without having to bother the Energy Board. Trunk Line might be concerned about getting "fair treatment" from the National Energy Board, but should gas consumers in the rest of Canada be asked to place their confidence in fair treatment by Trunk Line and its ability to determine what is "most economical . . . for all concerned?"

CLASHING PERSONALITIES

The corporate battle for a northern pipeline franchise almost immediately brought into conflict two strong personalities, both aggressive and ambitious, super-charged with energy and highly competent. They were Vernon Lyle Horte, who launched the first drive for a northern pipeline as president of TransCanada PipeLines and continued the pursuit as president of Arctic Gas, and Sidney Robert Blair, who broke ranks with the big guns of the industry to launch a guerilla campaign.

Images of each emerged from the scenes of battle. Horte, the articulate, pragmatic engineer, saw the challenge as designing a transportation system capable of overcoming obstacles of geography, economics and environment that lay in the path of distant but enormous sources of energy. His hope was that the most practical line would be politically acceptable. Blair, visionary and charismatic, searched for a plan that would be politically acceptable in the hope that it would be practical. Horte was tagged as the representative of the eastern establishment, clutching its traditional control of Canada's destiny, the front man for the multinationals in hot pursuit of mega-bucks. Blair was tagged as the western super nationalist ("a fanatic Alberta nationalist" according to *Toronto Star* columnist Richard Gwyn), the businessman with manure on his cowboy boots (newsmen were quick to report), out to wrestle control away from the hands of the multinationals and shift the power-base of Canada out west.

Blair has been described as a "businessman-politician," and there were published suggestions that he might enter the political field. For which party he would run is uncertain. He has said that Peter Lougheed "is western Conservatism at its best, and western Conservatism is often exactly the same thing as federal Liberalism."

Right from the start, Blair made Canadian ownership fundamental to his plans for a northern gas pipeline, and his statements have echoed an increasingly stronger nationalist view. "I'm in favour of foreign investment in Canada," Blair is quoted as having stated nine months after he joined Trunk Line. "I'm just even more in favour of Canadian investment." Nearly seven years later he told the National Energy Board that "the losses to Canada of allowing the imposition of what would be the biggest gas pipeline company in Canada under foreign ownership and management as proposed by Arctic Gas are simply too serious to be accepted. We will fight their proposition right into the ground with whatever time, money and enterprise we have." After the Energy Board had approved the Alaska Highway line in principle, Blair was quoted as saying: "The Toronto business establishment sees money and industrial power as ends in themselves and nationalism as a childlish emotion. To them, if it's right for Ford and Exxon, it must be right. Now we've proved that Canadians can stand up to Ford and Exxon."

Westerner Bob Blair is the grandson of an Alberta homesteader. His father, Sidney Martin Blair, retired on a farm near Toronto at age eighty-one in 1978, was an engineer with Bechtel Corporation of San Francisco. Bechtel is the world's largest firm of engineering contractors, building pipelines, petroleum refineries, railways, hydro electric projects, subways. Sid Blair rose through the ranks to become president of Canadian Bechtel, wholly-owned by the San Francisco firm. Bob Blair was born in Trinidad in 1929 where his father was the manager of a refinery. The family later moved to England, where Bob grew up, attending exclusive private schools there and later in Connecticut, before attending Queen's University in Kingston, Ontario. After graduating as a chemical engineer from Queen's in 1951, Blair worked for eighteen years with U.S. companies, first with Bechtel on construction of major pipeline projects in Canada, including the Trans Mountain oil line (built across the Rockies from Edmonton to Vancouver), Westcoast Transmission, and others. Blair moved to Alberta in 1959 to join the Calgary-based Alberta and Southern Gas Co. Ltd., a subsidiary of Pacific Gas and Electric Company of San Francisco, responsible for the gas supply that PG&E imports from Alberta. Starting as an operating engineer, Blair quickly rose in the organization to become vice-president in 1961 and president in 1966, three years before joining Trunk Line.

Journalists have suggested that it was, in part at least, his extensive experience with American-controlled firms that led to the full flowering of Blair's nationalism. In any event, he did not view his nationalism as any impediment to his quest for a northern pipeline franchise. "I knew the way opinion was running in Ottawa on Canadian ownership," he is quoted as stating following the National Energy Board decision. "I knew the government was keen to encourage and develop Canadian technology."

Vern Horte, seen as the eastern representative, is in reality typical of the legion of prairie farm boys who carved out professional and managerial careers in the petroleum industry, careers largely dependent, like Blair's, on the foreign capital that developed the industry. Born in Kingman, Alberta, Horte was raised on the farm of his Norwegian parents, and at eighteen joined the RCAF for the final two years of the Second World War. His veteran's credits took him to the University of Alberta where he graduated with a degree in Chemical Engineering. Horte started his professional career with a small Edmonton firm, Chemical and Geological Laboratories, worked for two years with the Alberta Energy Resources Conservation Board as a gas engineer, then five years in Dallas, Texas, with petroleum consulting engineers DeGolyer and MacNaughton. He returned to Alberta in 1957 to join TransCanada PipeLines, then being built, as chief gas supply engineer. In Calgary, Horte and Blair were weekend neighbors, with cottages at nearby Bragg Creek. In 1961, Horte was named vice-president, gas supply, with Trans-

Canada; in 1966 he moved to Toronto as group vice-president, and in 1968 was appointed president. Now a management consultant in Toronto ("I enjoy working on problems and I don't miss the administrative burden of a big organization," he says), Horte still has a farm near Kingman, which he visits often.

He was more relaxed than in the heat of the battle, but you could still raise a flash of Horte's old pugnacious style with the charges that Arctic Gas was run by the multinational oil companies, grasping to retain control of the gas pipeline business.

"It was a phony issue," Horte bristles. "Just look at the gas utility business in Canada. It's one of the few industries that is almost one hundred per cent Canadian-owned. Oil companies simply just don't want to invest in regulated public utilities, unless it's the only way that they can sell their gas. Just look at TransCanada. Sure, we built that with American money, and that's where we got most of the equity. There was nowhere else we could get it. And the only reason that the oil companies invested in that pipeline was because without it there was no way they could sell their gas. But what happened? Within three years, TransCanada was more than ninety per cent Canadian owned. Now we've got a seven billion dollar gas pipeline industry, it's one of the most important that there is to the country, it's Canadian-owned, and American money helped build it. I don't see anything wrong with using American money to do that, and I don't see anything wrong with doing it again, if we have to."

THE PROPOSALS EMERGE

The first announcement that plans were being studied for a pipeline to move Prudhoe Bay gas was made by TransCanada chairman James W. Kerr in June, 1969. Participating with TransCanada were Peoples Gas Light and Coke of Chicago and American Natural Gas Company of Detroit. It was an outgrowth of the studies, instigated nearly two years previously by Horte, of a line from the Pointed Mountain area. Kerr said that a line from Prudhoe Bay three thousand one hundred miles southeast to the Chicago area would likely have to be more than forty-eight inches in diameter to move the gas at a competitive cost, and that tremendous reserves, throughput volume and investment would be required. Cost would probably be "well in excess of one billion dollars." The feasibility studies were to be conducted by Williams Brothers Canada. Like

Canadian Bechtel, WBC was U.S.-owned, controlled by David Williams of Tulsa, Oklahoma, and like Canadian Bechtel it had been operating in Canada for many years, and was fully Canadian staffed. President of Williams Brothers Canada was a native Albertan, Phil Dau.

One week later, Westcoast Transmission and Canadian Bechtel announced plans for another pipeline proposal from the Alaskan North Slope to be built by Mountain Pacific Pipeline Ltd. Across the north, this line would follow the same route as that being studied by TransCanada, turning west at the southern edge of the Northwest Territories to follow the length of British Columbia to the U.S. border at Kingsgate, Idaho, where it would connect with the existing Alberta to San Francisco pipeline. Cost of the two-thousand one hundred mile Mountain Pacific line was estimated at $1.2 billion and completion was planned for 1973.

Vern Horte, meanwhile, was looking for more partners to join the TransCanada study, specifically the three North Slope oil companies, Exxon, Atlantic Richfield and Sohio. "It was apparent that this was going to have to involve a very major study, and we thought it would be advisable if the Alaskan producers could be convinced that it was in their interest to take a good look at this," Horte later told me. "We started talking to them in the fall of 1969 and we had many, many meetings with them over a period of months. We did a lot of preliminary work ourselves and then laid it out for them, trying to convince them this was something worthy of study. By late spring of 1970 they had decided that, yes, it was of interest to them."

In evidence presented to the Berger Inquiry, Blair said that Trunk Line in December 1969, the month he joined the firm:

conceived that it should be a part of any main project to transport gas from north to south across Alberta ... We were a natural link in the overall plan, so we asked if we could participate in the Northwest Project Study Group's organization [the group headed by TransCanada]. They replied no, that they would proceed with their studies and just might let us know later if there was any place for us in their project. This hit us very hard, because the important time in a project is in the period when it is conceived and shaped as to its future design and policies ... So within our company we concluded that we could not afford to be left out of such studies ... We

40

decided that we would advance an alternate study which we thought made more sense, both in engineering and public acceptance in Canada.

Horte says that he does not recall Trunk Line asking to join the studies at this time, but adds that they may have. In any event, after the three producers had joined and the six companies had then organized the Northwest Project Study Group, they did decide not to admit any new members at that time. According to Horte, "The group had decided that eventually any company that wanted to join would be welcomed. But at that stage we didn't even know if a pipeline would make any economic sense, and the group decided that things would move along faster if they didn't add any further members until they had a better handle on whether or not it looked like it was going to be economic."

Horte recalled that the group was approached by other U.S. gas pipeline firms, who were turned away and later joined the studies headed by Trunk Line. "I think maybe we played into that situation by not taking on additional members right off the bat," Horte said. "However, it was made very clear to all the companies that wanted to join that they would be able to do so at the appropriate time. It may have been that they should have let other members in sooner. However, I really don't think it would have changed the outcome. I think Alberta Gas Trunk Line, frankly, wanted to set up and establish a separate presence in any event."

When Blair did launch his own studies, he promptly made it clear that TransCanada could not join his club, either. The initial pipeline study efforts by Trunk Line were chronicled in a book called *People, Peregrines and Arctic Pipelines*, by Calgary writer Donald Peacock, subsidized by Trunk Line (although it did not have any say in what was written) and based on access to Trunk Line's files. Peacock records that Blair met with Horte and Trans-Canada vice-president George Woods (later president) in Toronto to outline the project which Blair then referred to as Trunk North. Horte asked whether TransCanada would be able to participate in the equity of the new pipeline company planned to carry northern gas. In a written report eight days later, Blair stated: "My reply discouraged that prospect, on the grounds that while TransCanada is a well-established and Canadian company, it would be better in principle to avoid any equity ownership or representation of any shipper if the transport service company is to maintain the complete autonomy to act as a neutral and common carrier, which we believe to be appropriate." That same principle did not appear to apply four years later when Blair was actively seeking the participation of TransCanada and other gas shippers to join Trunk Line in sponsoring Foothills Pipe Lines Limited.

Early in May, 1970, Blair met with Alberta Premier Harry Strom to outline the plans for Trunk North. Also present at the meeting were Mines and Minerals Minister A. R. Patrick, Attorney General Edgar Gerhard, Deputy Mines Minister Hubert H. Sommerville, and Conservation Board chairman Dr. George Govier. In a follow-up letter to the premier dated May 19, Blair referred to encouraging progress in "the important goal of obtaining and protecting a satisfactory position for the Alberta gas transport system within the overall development of the gas pipeline artery to be routed through western Canada from Alaska." He predicted that this line would become "the future main North American artery for natural gas transport."

Blair went on to state in the same letter that, "If the route of the first main passes centrally through Alberta, the Government and private citizens and companies in this province will be involved to a great extent and their public and private policies and investment desires will be reflected in the Alberta portion and throughout the entire development. Conversely, if the route passes by Alberta or crosses only a remote corner, the participation and voice of Alberta interests in the over-all development would probably be negligible."

"With so much of the capital investment in resource development being produced according to concepts and plans designed in distant headquarters," Blair wrote, "it would also be healthy to local morale and ambitions to mix in some leadership of large projects by local management, when such opportunity docs exist. There are real qualitative values in the example of a one and a half billion dollar enterprise in this markedly competitive field being conceived and run from within Alberta."

The first public announcement outlining the planned Trunk North project was made by Blair on June 29, 1970. It outlined a fifteen hundred and fifty mile system from Prudhoe Bay to tie into Trunk Line's facilities in Alberta, with ownership split into three sections. An American company would own the portion across Alaska; a federally-chartered Canadian company would own the section across the Yukon and Northwest Territories, while "the

Alberta section will be financed and operated by AGTL as part of its orderly and integrated expansion." After the project development period, AGTL would retain only a minority interest in the federal company that was to build the pipeline across northern Canada. The whole thing was to be completed in record time. Applications for approval to build the Trunk North system were to be filed that fall, the announcement said; construction would start in 1971, and the pipeline would be operational in 1974.

A few days later Mines Minister Russ Patrick issued a statement saying that the Alberta government was "delighted" with the Trunk North plans: "In view of Alberta's historic role in Canada as the predominant supplier of natural gas for Canadian and export markets, the Government of Alberta holds the strong view that any pipeline constructed for the transportation of natural gas from Alaska and northern Canada to Canadian and United States markets must pass through the province of Alberta and its producing territory." This clearly indicated what the Alberta government thought of the alternative proposals by the TransCanada group and Westcoast Transmission.

Two weeks after the Trunk North announcement, on July 15th, the Northwest Project Study Group was announced at a news conference held in Toronto by TransCanada and its partners. The six companies planned to spend two years on a $12 million research and feasibility study of the pipeline which they envisioned from Prudhoe Bay to Emerson and south to Chicago. The study was to be financed twenty per cent by TransCanada, fifteen per cent each by the two U.S. gas pipeline companies, and sixteen and two-thirds per cent each by the Alaskan oil companies, Sohio, Atlantic Richfield and Exxon.

"Neither our group nor anyone else is in a position now to say that a project of this scope and magnitude is feasible or financeable or to announce early construction plans," TransCanada chairman James W. Kerr stated. That was what the $12 million study was intended to determine; an obvious reference to the earlier Trunk Line statement which optimistically predicted a construction start the following year.

Northwest Project proposed a route from Prudhoe Bay to the American midwest that was three hundred miles shorter than Trunk Line's route through Alberta. The Northwest route, however, failed to provide a means to deliver any Alaskan gas to the California market, as Trunk Line would. At a news conference, Horte said that there were not enough gas reserves in Alaska to provide for a spur line to California. If and when adequate reserves were available, the existing line from Alberta to California could be extended north to connect with the line envisioned by Northwest. "Meanwhile, the U.S. midwest is the most logical and economic market for Alaskan gas," Horte said.

Canadian National Railway and the two major gas utilities serving the California market later joined with Trunk Line to form the Gas Arctic Systems (GAS) Study Group. For the next two years, Gas Arctic and the Northwest Project spent millions in studying their alternate proposals.

Chapter 4
The Uneasy Alliance

During a period of seven years, Bob Blair adroitly adopted a series of varying approaches in pursuit of the single objective of a controlling role for Trunk Line in the movement of any northern gas that crossed Alberta.

The first approach was to lead a consortium of companies which proposed to route Prudhoe Bay gas across Alberta and through the wholly-owned facilities of Trunk Line. Forced to merge into a larger group, Blair continued to fight for the use of the Alberta route and the Trunk Line system. When this group agreed to use the Alberta route but did not agree to use the Trunk Line system, another approach was developed. Trunk Line then proposed the "all-Canadian" Maple Leaf pipeline which would move only the Canadian gas from the Mackenzie Delta, leaving Alaskan gas to be moved by the "all-American" pipeline and tanker system proposed by El Paso. Finally, the Trunk Line approach evolved into two proposals: one pipeline to move Alaskan gas along the Alaska Highway route, and a second to move Delta gas.

For two years, starting in mid-1970, the Gas Arctic Systems group led by Trunk Line, and the Northwest Project Study Group developed separate proposals to pipe Alaskan gas across Canada, before they finally agreed to join forces.

From the outset, the six-member Northwest Project group appeared to be the front-runner. Led by Canada's largest gas pipeline and backed by three American oil companies, it seemed to have the inside track to the gas supply, and it had by far the more extensive research program.

Within weeks of the July, 1970, announcement of the program planned by Northwest Project, a mile of forty-eight-inch diameter pipe and a small spread of large pipeline construction equipment

45

and men were enroute to Sans Sault at the edge of the Mackenzie River, seventy miles northwest of Norman Wells and just south of the Arctic Circle, where a $3.5 million "Arctic Test Facility" was constructed. More than twenty varieties of grass were eventually grown over the buried pipe to determine which types would grow fastest (it grows amazingly fast with twenty-four-hour daylight during the short summer period), would best withstand the harsh winters and would provide the best insulation to protect the underlying permafrost.

An army of other consultants—engineers, metallurgists, economists, earth scientists, sociologists and biologists—were soon at work on a wide variety of tasks. Teams of biologists spread out in field camps across vast areas of the north engaged in the most extensive studies of northern wildlife that had ever been conducted. Parallel studies were soon underway by Trunk Line's Gas Arctic Systems Group, which built similar but smaller pipeline test facilities at Norman Wells and Prudhoe Bay.

These studies had barely started when the federal government announced "guidelines" for the construction and operation of planned northern oil and gas pipelines. Announced jointly by Energy Minister Joe Greene and Indian Affairs and Northern Development Minister Jean Chretien on August 13, 1970, the guidelines described the planned pipelines as offering a "potential major economic contribution" to the country. The guidelines provided that "initially one trunk oil pipeline and one trunk gas pipeline will be permitted to be constructed in the north within a 'corridor' to be located and reserved following consultation with industry and other interested groups." Applicants for a pipeline permit were required to provide "substantial opportunity" for Canadian participation in financing, engineering, construction, ownership and management; document studies on the environmental effects of proposed pipelines and plans to mitigate adverse effects; and provide programs for the training and employment of northern residents on pipeline construction and operation. An expanded version of the guidelines was later tabled in the House of Commons on June 28, 1972 by Chretien. The guidelines clearly indicated at least some of the major areas of study which any application for a northern pipeline would have to undertake.

One of Blair's first tasks had been to seek other participants for the Trunk North studies. In September 1970, Canadian National Railways vice-president Dr. R. A. Bandeen (who is Blair's brother-in-law) announced that the CNR had been working with Trunk Line on the initial studies since June, and later signed a more formal participation agreement. In August, Northern Natural Gas Company of Omaha announced that it would join Trunk Line in the studies. On December 11, 1971, Blair announced that two other U.S. firms had joined, and the group now became known as Gas Arctic Systems Study Group. The two newest members, bringing the total to five, were Columbia Gas Systems, Incorporated of Willimington, Deleware, and Texas Eastern Transmission Corporation of Houston, Texas.

Westcoast Transmission, meanwhile, was working on its plans for Mountain Pacific Pipeline Ltd. In November, 1970, Westcoast announced that, in addition to Canadian Bechtel, three other firms had joined this group. They were El Paso Natural Gas of El Paso, Texas, and two Los Angeles gas utilities serving southern California, Pacific Lighting Service Company and Southern California Edison Company. But Mountain Pacific never did get off the ground. By September, 1971, Pacific Lighting had left the Mountain Pacific group to join Gas Arctic. Later, El Paso was to promote its own plans for an "all-American" system to ship Alaskan gas by pipeline and tanker.

Northwest Project established offices in Calgary in October, 1970, with the appointment of Lee Hurd as general manager. Hurd, a native Albertan with eighteen years experience as a gas pipeline engineer, had been vice-president of Alberta and Southern Gas Co. Ltd. when Blair was president of that firm.

With the research and studies being conducted by Williams Brothers Canada and a host of other consultants, Northwest Project (later Arctic Gas) had a total staff of only five in May, 1971, when I left my job as editor of *Oilweek* to become director of public affairs for this group. Six years and three months later I left Arctic Gas, just as the office furniture was being auctioned off.

THE URGE TO MERGE

Participants in each of the competing proposals were under no illusions that ultimately they would nearly all have to join forces if a pipeline were to be built to move Canadian and American gas from the western Arctic. It was not the type of competition where a single company could win the franchise and do the job by itself.

"It seemed ridiculous for two groups to be competing with one another for a project that was so large that fundamentally it had to be an international and industry-wide project," Vern Horte said in an interview in late 1977. "Nobody should be left out, it was going to be that big."

Blair and Horte began discussions on the possibility of merging the two groups in August, 1971. Representatives from the two groups continued discussions in a series of meetings in Calgary, Toronto, Houston, Cleveland and Omaha for the next ten months.

Major points at issue in the discussions involved Canadian ownership, routing, and the use of Alberta Gas Trunk Line's system in Alberta. Both groups agreed that Canadian ownership must be an initial objective. The Northwest companies, however, were not convinced that it would be possible to initially raise more than fifty per cent of the ownership capital in Canada and expressed concern that a firm commitment to this might make it impossible to finance the undertaking. Blair felt that the majority, if not all, of the equity could be raised in Canada at the outset, and that there should be a firm commitment to this. On routing, TransCanada and the two American midwest gas companies favoured the Northwest Project route, since it was a more direct and shorter route to the markets in Ontario and the southern midwest area. Trunk Line wanted a route down the length of Alberta. The final major point was whether Trunk Line would build and operate the portion of the system in Alberta.

According to Horte, the initiative in these discussions came from the Northwest group. "Blair was extremely reluctant all the way through," Horte said. "At meeting after meeting, it was an extremely frustrating experience. Continually he [Blair] would give lip service to wanting to get together, and on the other hand I came away with the feeling that unless it was put together exactly the way he wanted it, which was fundamentally that he wanted to control the thing, or wanted the Canadian participants basically to have the complete say as to how the studies, etc. were to be carried out. The U.S. participants would simply be kept informed, but they would just keep putting up their share of the money...It was really absurd to think that the U.S. companies were going to agree to put up a great deal of money and have no say, but would simply be kept reasonably informed."

Blair has expressed a different view. He has claimed that the Northwest group was dominated by U.S. companies, motivated to meet U.S. interests at the expense of Canadian and Albertan interests.

Horte and Blair appear to agree on one point, however. It was pressure from the U.S. partners in the Gas Arctic Systems group which in the end brought Trunk Line into the eventual merger.

"We would never have made a deal, in my opinion, with Blair," Horte said. "What we had to do was broaden the negotiations beyond just Blair to bring in the other companies to make sure they heard the whole thing." Horte says that at the final negotiations, the U.S. participants in Gas Arctic "were prepared to disband their interest in that project and leave Blair all on his own. They were going to back out and join our project in any event, because they felt it had done more work, and that there shouldn't be two groups fighting one another."

In Peacock's book, Blair is quoted as stating that in the merger, Trunk Line "accepted terms we otherwise wouldn't have because of pressure from our American partners."

The negotiations had progressed far enough that at a meeting in Omaha the twelve companies involved had agreed that three people would temporarily serve as co-chairman of the merged group. They were Blair, Horte and W. H. (Deke) Mack, president of Michigan Wisconsin Pipe Line Company, the major subsidiary of American Natural Gas Company of Detroit.

The twelve companies in the Northwest and Gas Artic groups, plus four new firms, were finally betrothed in a group marriage which took place June 8, 1972, in the Sonesta Hotel in Houston. Discussions started with breakfast at 7:30 and lasted all day before the sixteen firms finally agreed to the terms of a thirty page "Joint Research and Feasibility Agreement." Horte acted as chairman at the meeting, at which there were some fifty people.

The agreement, dated June 1, recognized that the contemplated pipeline would "involve the expenditure of extremely large sums of money ... novel financing plans" and "novel technical aspects and regulatory and other governmental proceedings and authorizations. The immensity and unique nature of the project indicate that the cooperation and participation of a substantial segment of the natural gas industry will be desireable, and may be essential, to the successful completion of the project." To achieve such cooperation and "conform to the guidelines published by the Canadian government," the firms "agreed to consolidate their future activities and past studies relative to the gas pipeline project." They also agreed to "promptly seek as participants other Canadian and United States companies who have an interest in the project and whose participation may contribute to the objectives of the project." All of this, it was said, was in recognition "of the pressing need of Canadian and United States consumers for additional supplies of natural gas, and the certainty that such need will become increasingly more severe, which requires all feasible expedition of the subject gas pipeline project."

Several features of the agreement were key to its acceptance. These included a commitment to take a fresh look at all possible routes, as well as the use of existing pipelines: "...the Study Group shall study and consider all reasonably feasible gas pipeline configurations, routes and facilities and methods of ownership ... including wholly new facilities and the utilization of the whole or any portion of any presently existing systems as it may now be or as it may be expanded or otherwise adapted."

Another provision was for maximum "feasible" Canadian ownership. Canadian-owned firms would "be given the prior opportunity" to invest in any corporation owning the planned facilities within Canada, while "Canadians shall be given the opportunity to acquire ownership in the corporation to the maximum extent feasible and consistent with the formulation of a practicable over-all permanent financing plan." The only limitation to Canadian ownership would be the amount of money that Canadian firms and individuals might choose to invest in the pipeline. But no firm commitment to majority Canadian ownership was spelled out in the agreement.

A final item, the provisions for voting procedures, was designed to provide the Canadian-owned firms with a degree of control during the study phase, even though they represented a minority of the members in the group. The four new member firms included one Canadian-owned firm (Canadian Pacific Investments) and the Canadian subsidiary of three multinational oil companies active in Mackenzie Delta exploration: Imperial Oil, Shell Canada and Gulf Oil Canada. With these new members, only four of the sixteen firms were Canadian-owned: TransCanada, Alberta Gas Trunk Line, CNR and CPI. Expenses of the group, which eventually exceeded $150 million, were divided evenly among all member companies.

The voting procedure split the members into three categories, essentially the Canadian-owned firms, the United States pipeline firms, and the oil companies. Major decisions by the group required an affirmative two-thirds vote by each category. The effect was to give the Canadian-owned firms a veto, even though they put up only one quarter of the money.

The combined group was known as the Gas Arctic-Northwest Project Study Group. Policy decisions were in the hands of a management committee, comprising representatives from each of the participating companies. The management committee normally met monthly, usually in Toronto. The study group organized two principal companies, Canadian Arctic Gas Pipeline Limited and Alaskan Arctic Gas Pipeline Company, which were intended to conduct the studies, pursue the regulatory applications, and, if successful, finance, build, own and operate the pipeline. At the time of financing, the management committee would be phased out, to be replaced by a conventional board of directors.

A new engineering organization, Northern Engineering Service, was organized to act as primary consulting engineers, responsible for the technical studies. NES was owned seventy-five per cent by five Canadian engineering firms and twenty-five per cent by Resource Science Corporation of Tulsa, Oklahoma, parent company of Williams Brothers Canada.

Based in Calgary, NES developed into one of the most extensive pools of technical expertise ever assembled in Canada. It conducted close to $100 million in studies and research. At its peak, NES employed one hundred and seventy people and engaged the services of more than thirty other consulting organizations, ranging from tiny Golden West Seeds of Calgary to giant Canadian Pacific of Montreal. It was exactly the type of technical resource base that the government was so eager to see developed, and which takes years to duplicate. NES was disbanded, and this technical resource base shattered, following the government's 1977 pipeline decision. Foothills could not retain this group of experts to work on the Alaska Highway pipeline proposal, because it did not have the budget to keep such a large technical organization usefully engaged. Phil Dau, president of NES, had spent ten years directing northern gas pipeline studies. The last time I saw Phil was in September, 1977, in the derelict offices of NES. He was arranging the disposal of the company's remaining assets: some electric typewriters, a tape recorder, a slide projector, half a dozen Arctic parkas.

51

Ten days after the merger meeting in Houston, William Price Wilder, fifty, president of Canada's largest investment firm, Wood Gundy Limited, Toronto, was appointed chairman and chief executive officer of the Gas Arctic—Northwest Project Study Group and the Arctic Gas companies.

Wilder's entire professional career had been with Wood Gundy where he played an active role in arranging the financing for some of Canada's largest undertakings, among them TransCanada Pipe-Lines and the Churchill Falls hydroelectric project. Wilder's father, also a William, had been a vice-president of Wood Gundy, who had died suddenly at age forty in the Spring of 1929. Before his death, the senior Wilder had borrowed a million dollars from the bank to invest heavily in shares of Massey-Harris, at prices of up to $96 each, later falling to $5 in the 1929 stock market and wiping out an estate that had been estimated at between four and five million dollars. Wilder was still able to attend Upper Canada College, the Toronto establishment school whose enrollment in 1976 included the sons of both Bill Wilder and Ontario NDP leader Stephen Lewis.

Wilder completed Upper Canada College just in time to spend four years in the Second World War where he served as a lieutenant with the RCNVR on loan to the Royal Navy. After the war and a Bachelor of Commerce degree from McGill University, Wilder joined Wood Gundy as a security salesman in 1946, later attending the Harvard School of Business Administration. By 1961 he had risen to executive vice-president of Wood Gundy, and in 1967 was appointed president.

Stepping from the comfortable obscurity of the financial community into a high profile role at controversial Arctic Gas, Wilder faced a new role as a public spokesman, a quasi-political role. It was not a role with which he was naturally comfortable. He tackled it with the same dedicated determination that he had tackled every other career challenge. Eric Kierans, the Montreal economist, former federal cabinet minister and early critic of the Arctic Gas project, called Wilder a "corporate circus barker." Wilder brought to Arctic Gas not only astute business judgment, but even more importantly the solid reputation of trust and integrity that he had built in twenty-six years at Wood Gundy, and a blunt candor that at times verged on the undiplomatic. On any given question, no one has ever had to wonder where Bill Wilder stands.

The next major appointment at Arctic Gas was announced two weeks later when Vern Horte was named president. Wilder and Horte both assumed their positions at Arctic Gas September 1, 1972.

From the sixteen members at the time of the merger, the number of participants in the Arctic Gas group quickly increased to a peak of twenty-eight. Among the new members were the natural gas distribution utilities serving most of the Canadian markets: Canadian Utilities Limited (serving the major Alberta centres), Consumers' Gas Company and Union Gas Company (serving southern Ontario), Northern and Central Gas Corporation and its affiliates (serving markets in Manitoba, northern Ontario and Quebec).

One by one, nearly half the companies dropped out. The American oil companies had made it plain from the start that they were only in until the project reached the regulatory stage. Others dropped out because, with Arctic Gas spending $30 million a year, it had become too expensive for them to stay. Only one company, Trunk Line, dropped out of the group in order to pursue an alternative proposal.

During the months after the merger, Horte and Wilder assembled a powerhouse management group at Arctic Gas. John Yarnell, a native Montrealer who was raised in Winnipeg, came from Consolidated Bathurst Limited where he had been group vice-president to become vice-president in charge of finance. Yarnell had earlier worked with Gulf Oil Canada, where he had been corporate treasurer. Jim Harvie, a native Albertan raised in the oil fields at Turner Valley, was hired from Gulf Oil Canada to head engineering and operations in Calgary. In Ottawa, Bob Beattie, former Senior Deputy Governor with the Bank of Canada, was retained as an economic consultant. Bob Ward, former Lieutenant Governor of Alaska, was hired as president of the Alaskan Arctic Gas Pipeline Company with offices in Anchorage and Washington. Chicago lawyer Bill Brackett moved to Toronto to head the legal side as vice-president of corporate affairs, later transferred to Washington as vice-chairman of Alaskan Arctic Gas. Ted Creber served for one year as vice-president and general counsel before leaving to become president of The Consumers' Gas Company. The position of general counsel was then handled by Harry Macdonell of the Toronto law firm of McCarthy and McCarthy, the firm which former federal Finance Minister Donald Macdonald joined in late 1977. Macdonnell was also the former president of the billion-dollar Churchill Falls hydroelectric project in Labrador.

IN SEARCH OF A ROUTE

In his first public speech as Arctic Gas president, Horte outlined where the project stood three months after the merger agreement. It was hoped that applications could be filed by mid-1973, and if early approvals were obtained, the pipeline could be in operation by 1978. Alternate routes were being studied "to determine the best and most economic means of moving gas to the major market areas in both Canada and the United States. Hopefully, by the end of this year we will have our routing plans defined." In fact, it was to take more than another year of dispute and wrangling before the member companies could agree on a route and on how the pipeline should be owned, decisions which eventually led to Trunk Line's withdrawal to promote its own plans.

One of the first tasks facing Northern Engineering Services was a new study of routing alternatives. The NES study, completed that October, examined three different primary routes; it encompassed six cases based on different assumptions about the volumes of gas in different regions. Bound in red covers, this study became known in the group as the "red book".

One week later, Blair telexed the member companies stating that the terms of the merger "call for the consideration of expansion of existing systems as one of the alternatives for transportation of arctic gas," and claiming that the red book study "does not provide information on this alternative. We believe it would be improper and not in accordance with terms of the study group agreement to make any route selection until the possibility of expansion of existing systems is fully considered." Blair advised that Trunk Line and TransCanada were conducting their own studies. Known as the "green book" study it examined four cases and was presented to the member companies on November 7. Blair made his interests clear in a speech delivered November 17, 1972 to the Institutional Investors Conference in Toronto. He said Trunk Line would "have a direct business interest in investing in over one billion dollars of additional pipeline plant integrated with our present system across Alberta" if expansion of existing pipelines were used as a means of moving northern gas from south of the Northwest Territories. He

54

said that, "various alternatives are being considered towards selection of the most advantageous means of transporting the arctic gas south of the 60th parallel of latitude. One of the alternatives which is receiving consideration is expansion of the existing Canadian pipeline system." In addition to expanding its own system, Blair said that Trunk Line "would consider putting as much as one hundred million dollars" equity investment into a new Canadian company financed to "provide transportation services at least across the Yukon and Northwest Territories."

At the group management committee meeting in Toronto on December 7, 1972 there was general agreement that the system proposed by Trunk Line and TransCanada would involve no increase in capital cost or cost of service for the initial volumes of gas that were contemplated. At projected greater volumes, however, there would be a cost penalty for consumers in Ontario and the U.S. midwest. There was also concern expressed about segmenting the ownership of the pipeline in Canada. It was argued that a single company could better plan and co-ordinate the undertaking, avoiding the alleged risk of a chaotic situation with different companies bidding against one another for services of contractors, construction equipment, material and crews. There was also concern expressed about whether Trunk Line could, by itself, finance the estimated one and a quarter billion dollars in new pipeline construction that would be required in Alberta.

The minutes of that meeting record: "For Alberta Gas Trunk Line, Mr. Blair expressed the belief that further study would reveal a greater economic advantage in favour of the TransCanada and AGTL proposals. In any event, however, he maintained that the latter involved important non-cost advantages, which should be given considerable prominence in the making of the final decision."

Management of Arctic Gas was requested to prepare its assessment for consideration by the member companies. In their report presented to the member companies on May 9, 1973, the Wilder-Horte team addressed both the routing and ownership questions. On ownership, they quoted the findings of the financial advisors, Wood Gundy and Morgan Stanley & Co., who concluded "that multiple ownership of facilities would have serious adverse consequences on the form and feasibility of the project's ultimate financing."

"We have made our recommendations," the staff report stated, "in the conviction that under any form of ownership under consideration, the elements of Canadian direction and content of the project will be at an acceptable level: CAGPL will be Canadian controlled and will be operated in accordance with Canadian law and regulatory policy. Further, we believe that a commonly owned Arctic Gas enterprise, like the separate entities proposed, would be sensitive to and comply with local governmental and public concerns throughout Canada."

On the routing question, the report observed: "A very important consideration underlying our recommendation is our conclusion that the system built should, as far as possible, be able to make use of excess capacity on existing pipeline systems, if such capacity, not presently available, comes into being."

In short, the staff report sided with Trunk Line on the routing and on the use of excess capacity in Trunk Line's system, but opposed Trunk Line's proposal for separate ownership of that portion of the line in Alberta.

Some further problems with the separate ownership idea were spelled out in a report prepared by Trunk Line's lawyers, Campbell, Godfrey and Lewtas of Toronto, on "Legal Considerations Relating to Routing Alternatives and Ownership." The lawyers suggested that Trunk Line may have been built under the "wrong jurisdiction," and, "whether AGTL's present system is constitutionally a provincial or a federal undertaking, the use of the system to carry arctic gas through the province to extra-provincial markets would result in at least that portion of the system carrying arctic gas being part of an inter-provincial system and subject to federal jurisdiction." The lawyer's report noted that Trunk Line proposed to meet this transmission problem by creating a federally-incorporated company, "in which AGTL would have a significant share interest." The report also mentioned legal speculation that "even as presently constituted and operated it [Trunk Line] is constitutionally subject to federal rather than provincial authority. If this be the case, a number of difficult questions arise as to the consequences of having acquired, constructed and operated its existing facilities under the laws and regulations of the wrong jurisdiction. So long as this uncertainty persists it would be necessary to consider its impact on potential investors in the project."

The report continued: "The cases which do not involve the incremental looping of existing systems avoid the complications resulting from contact with existing systems, and therefore offer the greatest simplicity and flexibility of legal structures and the fewest constraints. Any case involving incremental expansion of existing facilities, since it requires the use of AGTL facilities to carry arctic gas, raises difficult questions of constitutional law and of the interpretation of the BNA Act." But the report ended on an optimistic note: "There is nevertheless no reason to believe that the issues may not be appropriately resolved." Just how this would be accomplished was not suggested.

The first indication of a break in the deadlock over the ownership question appeared at the management committee meeting on May 30, 1973 when TransCanada chairman James Kerr presented a statement saying that his company was prepared to accept the single ownership concept provided that certain conditions were met:

1) The routing of the project will follow existing corridors and pipeline routes.
2) A policy of incremental expansion and common use of existing pipeline facilities will be followed where practical.
3) Contractual guarantees are given to Alberta Gas Trunk Line and TransCanada that their facilities will be kept fully loaded at all times and under all conditions.
4) While existing facilities are to be kept fully loaded, Trans-Canada would expand its facilities to the extent additional volumes of gas in Alberta and/or ASaskatchewan are acquired.

Minutes of the meeting record that "in the ensuing discussion, every participant except Alberta Gas Trunk Line, Pembina and Columbia expressed its support of the single ownership question." (Pembina Pipelines and Columbia Natural Gas did not state a position.)

There was, however, little acceptance of the TransCanada stipulation that would have obliged Arctic Gas to keep the trunk Line and TransCanada systems "fully loaded at all times and under all conditions." This could have obliged the Arctic Gas pipeline to operate at less than capacity so that Trunk Line and TransCanada's facilities would. In the view of several member companies such a stipulation would make it impossible to secure financing even in the unlikely event of approval of this feature by regulatory authorities.

There were more studies, more alternatives, more compromises, before a solution was unanimously approved at the management committee meeting on June 27—more than a year after the merger. The agreement provided that all gas destined for Canadian markets east of Alberta would be transported from the Alberta border by TransCanada; that American gas for the midwest market would be picked up by a new pipeline on the Saskatchewan-Montana border at Monchy; that "single ownership of a complete, integral pipeline system" would extend across Alberta, following the route of the Trunk Line system. It also stated: "After completion of the initial Arctic Gas system, full consideration will be given to the use of any long-term unused capacity if economically available in the AGTL or ANG systems as a preferred alternate to direct expansion, provided undue engineering or operating problems are not thereby induced."

With this agreement finally in hand, Arctic Gas was then able to continue with the task of preparing thousands of pages of supporting evidence which would be filed when applications were filed simultaneously in Ottawa and Washington on March 21, 1974.

Even before the applications were filed, dramatic events in the Middle East had led to strong statements of public support for the Arctic Gas project by the Government of Canada. On November 23, 1973, Prime Minister Trudeau addressed the nation on television to discuss the new-found "energy crisis" precipitated by the six-day war in the Middle East and the embargo of Arab oil supplies to western nations. He talked about "measures the Canadian government must take to deal with a serious problem of oil supplies which we, like other nations, face this winter." Government and industry, Trudeau said, are "racing against the onset of winter to move all available Canadian and foreign oil from wellheads and refineries to the homes and industries and institutions where it is most needed," by pipeline, railway, truck, and by tanker shipments from Vancouver through the Panama Canal to the Atlantic provinces. He promised that he would later discuss "more fully Canada's future energy concerns and how we are dealing with them." He promised decisions answering "questions about a national energy company and about construction of a Mackenzie Valley gas pipeline."

In the House of Commons on December 6, Trudeau spelled out the additional details: "A major development is the proposed gas pipeline up the Mackenzie Valley to move Alaska gas to the U.S. markets and at the same time make it possible to move Canadian northern gas to Canadian markets. While this project must, of course, be subjected to the usual regulatory proceedings and cannot go ahead until it has been approved by responsible Canadian authorities, the Government believes it will be in the public interest to facilitate early construction by any means which do not require the lowering of environmental standards or a neglect of Indian rights and interests. At this point, I might just say that I can see no reason why Canada could not give suitable undertakings as to the movement without any discriminatory impediment, of Alaskan gas through the pipeline across Canada to U.S. markets, provided all public interest and regulatory conditions are met in the building and operation of the pipeline. An undertaking of this sort would, of course, be reciprocal, with the same assurance being given Canada regarding our own oil and gas shipments through the United States."

Canada survived the winter, the Arabs resumed their oil shipments (at quadrupled prices), and in the warmth of spring any national sense of urgency about pipeline construction had faded like the morning dew. Two people in particular who were not wholly enamored with Arctic Gas plans for a Mackenzie Valley gas pipeline were Kelly H. Gibson, chairman of Westcoast Transmission, and Bob Blair.

Abandoned by all its partners in the Mountain Pacific project, Westcoast had declined invitations to join the Arctic Gas group; it was the only large segment of the Canadian gas industry that was not participating. On April 23, 1973, Gibson hinted to the company's annual meeting of shareholders in Vancouver that Westcoast might still have plans of its own. Gibson predicted that the regulatory proceedings confronting Arctic Gas will require "several years," that it "will become a matter of great public debate," and that "there are a number of alternatives to this ambitious plan." Westcoast had been studying "this entire question" for at least five years, he assured the shareholders. "The company's destiny will be affected by the transportation and sale of frontier natural gas, by whatever means or route. At the appropriate time, Westcoast will enter this important debate in an informed and constructive way."

Amplifying these alternatives to newsmen, Gibson said that they "don't need Alaskan gas." This could only mean that he was already thinking of a pipeline to transport only the gas that had been discovered in the Mackenzie Delta.

Six days before Gibson's Vancouver remarks, Blair had written to Alberta Mines and Minerals Minister W. D. Dickie to express his distress over the fact that the Arctic Gas project did not conform to the corporate interests of Trunk Line. "In our company, we are distressed with the trend of events connected with the filing of the Arctic Gas pipeline application by Canadian Arctic Gas Pipeline Limited," Blair wrote. "Our distress is in two directions: first, we do not believe that the application, as filed, by the majority vote of the 27 members, is sufficiently tailored to meet the future commercial involvement of our company or of gas services in and for the province as reflected through our operations. Second, we judge that the project applied for may not be a practical proposal when assessed in the whole range of considerations covering ownership, financing, service to Canadian markets and political acceptability in Canada."

It was only a matter of time before Trunk Line and Westcoast would join forces to promote an alternative plan. This, in time, became the "Maple Leaf" project of Foothills Pipe Lines Limited, designed to transport Mackenzie Delta gas to expanded facilities of Trunk Line and Westcoast. Events moved in this direction over a period of several months in 1974.

Blair was already studying plans for a forty-two-inch diameter pipeline to move Delta gas, and advised Arctic Gas member companies of these studies in a prepared statement presented to the management committee meeting on May 22, 1974. According to his report, the proposed pipeline would cost $1.75 billion and could be in operation by 1979. These studies, he explained, were undertaken as a "contingency plan" in the event that Arctic Gas lost out on the competitive bid by El Paso to move Alaskan gas, or if it were otherwise delayed in the task of providing Delta gas in time to meet the urgent need of Canadian gas consumers. Blair claimed that this proposal offered the following advantages: it was not dependent on American approvals or on the availability of Alaskan gas; it would be easier to finance than the larger system proposed by Arctic Gas; it would be rationalized with existing pipeline systems; it would avoid uncertainties associated with the use of forty-eight-inch diameter pipe; and it would be more responsive to the objectives of provincial governments. However, the memo also stated that while studying this contingency plan Trunk Line would continue to support the Arctic Gas project.

"Alberta Gas Trunk Line is fully supporting the Canadian Arctic Gas application," the Toronto newsletter *Energy Analects* reported two weeks later, " . . . but should CAGPL fail to win approval from Canadian and U.S. regulatory authorities, AGTL would be quite prepared to see to it that Delta gas reaches Canadian markets." Asked if Trunk Line were thinking of leaving Arctic Gas, *Energy Analects* reported that Blair "gave a firm no."

On July 31, Blair held a news conference in Toronto at which he issued a six-page statement marked for release "immediately after the management committee meeting of the Gas Arctic—Northwest Project Study Group," which was being held that day in the Royal York Hotel. "The Alberta Gas Trunk Line Company Limited continues to be a member of the Gas Arctic—Northwest Project Study Group and to share costs and results," the statement read. It went on to outline, in greater detail than the May 22nd statement, the same basic advantages claimed for "an alternative or contingency plan for the construction of a natural gas pipeline from the Mackenzie Delta—Beaufort Sea area to Canadian markets." The statement stressed Canada's need for new gas supplies: " . . . every single gas requirement forecast we have made or seen shows a shortfall by 1979 or 1980 in the ability of Alberta, British Columbia and Saskatchewan to meet the growing requirements for the whole country." While the contingency pipeline was predicated on the movement of only Delta gas, it "could also be expanded readily to carry Alaska gas as and when such services may be sought."

Bill Wilder issued a one-page news statement the same day. He said it would be "wasteful and folly if Canada built a natural gas pipeline from the Mackenzie Delta and the U.S. built one across Alaska, without seriously and strenuously working towards a joint system."

In Toronto on August 27th, Trunk Line officials, led by counsel Reg Gibbs, met with representatives of four Canadian member firms of Arctic Gas (TransCanada, Consumers' Gas Company, Union Gas Limited, and Northern and Central Gas Corporation) to seek support for Foothills' Maple Leaf project. According to a written report of the meeting by one of the participants, Gibbs "described the Canadian supply position by 1977 as precarious." Gibbs suggested that "there is an obligation on Canadian transmission and distribution companies to get behind the AGTL project to bring Delta gas to Canadian markets. Gibbs said that Foothills Pipe

Lines (the entity that would operate north of 60°) would come under federal jurisdiction, but that they were hopeful all pipelines operating in Alberta would be subject to regulation only by Alberta authorities. He admitted there was some question as to whether such a split was constitutionally possible and said that if ordered to do so the AGTL facilities involved in eastern Canada deliveries would also be regulated by the National Energy Board." None of the Canadian companies left Arctic Gas to join the Foothills project.

On September 13, Trunk Line notified the other companies in the Arctic Gas group of its intention to withdraw. It was not, by that time, an unexpected move. Three days later, Trunk Line announced that it was being joined by Westcoast Transmission in the ownership of Foothills. Kelly Gibson was named chairman of Foothills and Blair, president and chief executive officer. Only one other company—Northwest Pipeline of Salt Lake City, Utah—had joined the Foothills effort before the Arctic Gas project was rejected three years later.

Quick support for the Foothills project was voiced, however, by Alberta Premier Peter Lougheed. On September 18, *The Calgary Herald* reported: "Premier Peter Lougheed said today the proposed Maple Leaf line to carry gas from the Mackenzie Delta 'would be preferable for Canada' to the Canadian Arctic Gas Pipeline international proposal. Mr. Lougheed said in a telephone interview from Edmonton that his 'government would tend to give the Maple Leaf proposal preference,' possibly including financial backing if necessary."

In later testimony to the Berger inquiry, Blair recalled his reason for Trunk Line's withdrawal from Arctic Gas to promote its own proposal, " ... the root of Alberta Gas Trunk's decision to withdraw from the Arctic Gas group was our judgment that Arctic Gas was pursuing too much the purpose of the original Northwest Project Study Group's desires to create an internationally known and wholly new express line across western Canada, and was doing so too much for the special purpose and under the influence of the United States companies."

Chapter 5
The Hearing That Never Was

It seemed astonishing that the chairman of the National Energy Board would be found unqualified to sit on the panel conducting what were considered the most important hearings the board had ever held.

Marshall Crowe was not only one of the most highly regarded of the mandarins in the Ottawa civil service, but he had also held some important positions in the business world. When he joined the board to become its chairman in late 1973, it was with a full sense of the importance of the challenge that lay ahead. Decisions on energy matters confronting the board would be crucial to the future of Canada, and none seemed more so than the decision that would have to be made on a Mackenzie Valley gas pipeline. A key role in this decision would climax his career in serving the public interest.

Yet, two years and five months after he had joined the board, here was the Supreme Court of Canada ruling that Marshall Crowe was unqualified to hear the competing applications of Arctic Gas and Foothills Pipe Lines. "The participation of Mr. Crowe in the discussions and decisions leading to the application made by Canadian Arctic Gas Pipeline Limited . . . in my opinion cannot but give rise to a reasonable apprehension, which reasonably well-informed persons could properly have, of a biased appraisal and judgment of the issues to be determined," Chief Justice Bora Laskin had written, outlining the court's majority decision. Five of the court's justices had concurred in the decision and three had dissented, holding that there was no impediment to Crowe conducting the public hearings.

The court's decision came on March 11, 1976, an awkward time. It had been two years since Arctic Gas had filed its applications, and the National Energy Board was lagging well behind the other public inquiries into this matter. In Yellowknife, the Mackenzie

Valley Pipeline Inquiry headed by Justice Berger had been in progress for a full year. In Washington, hearings before the Federal Power Commission on the competing applications of Arctic Gas and El Paso had been in progress for ten months. The National Energy Board had been the last to get its hearings started, eight months after Justice Berger and six months after the FPC. And now, because of the ruling by the Supreme Court, the board would have to start its hearings over again, under a new hearing panel of different board members.

Apprehensions about possible bias arose from the fact that prior to joining the board, Crowe had been president of the government-owned Canada Development Corporation. CDC was a member of the Arctic Gas group of companies, and Crowe had served for one year as a member of the Arctic Gas management committee, which is comparable to sitting on a board of directors. He had participated in some of the key decisions which Arctic Gas had made leading to its application to the board.

In the legal proceedings which had removed Crowe from the panel of three NEB members hearing the applications, there was never any suggestion that he was biased, in fact, or had acted in any way improper. The legal issue was whether or not anyone might suspect that he would be biased because of his prior association with Arctic Gas. "The central issue in this case," in Chief Justice Laskins' words, was "whether the presiding member of a panel hearing an application...can be said to be free from any reasonable apprehension of bias on his part when he had a hand in developing and approving important underpinings of the very application which eventually was brought before the panel." The Chief Justice expressed a "firm concern that there be no lack of public confidence in the impartiality of the adjudicative agencies."

If it seemed astonishing to some that the chairman could not participate in the most important hearings conducted by the board, the court's decision at least was not a complete surprise. More than two years before, a legal firm had advised Arctic Gas that if Crowe sat on the panel hearing the applications there was a risk that this would happen. And Crowe, at the time, had been apprised of this legal opinion.

A grain of worry had started to rub in the back of the legal minds at the offices of Arctic Gas when Crowe's appointment to the National Energy Board was announced in October, 1973. If Crowe were to chair the hearings, there was a risk that any decision favourable to Arctic Gas might be attacked in court, and both the hearings and decision declared null and void. The whole hearing exercise would have then been a waste of time and money, as much as two years and millions of dollars. There was an even greater danger. With El Paso already having filed its competitive all-American pipeline and tanker proposal with the Federal Power Commission in Washington, such a delay could well prove fatal to the Arctic Gas proposal.

There were other pressing concerns facing Arctic Gas and its twenty-eight member companies as they prepared to file applications with the Canadian and United States authorities. One of the issues was brought into sharp focus by the first company to withdraw from the group, Sohio.

On March 7, 1974, two weeks before the applications were filed, Sohio vice-president Earl Unruh wrote to Arctic Gas and the member companies that the time had come for Sohio to drop out. Sohio, as well as the other two American oil companies, Exxon and Atlantic Richfield, had made it clear from the outset that they intended to support only the research and studies and had no intention of participating in the ownership of any subsequent gas pipeline. With the applications about to be filed, Sohio had concluded that this was the time to get out, and it was not too long before Exxon and Atlantic Richfield also withdrew.

In withdrawing, Sohio raised once again an issue on which there was as yet no agreement among the member companies: could the pipeline be built without financial support from the government? Sohio thought not.

"Sohio feels very strongly that in order to make the project fully viable, both Canadian and U.S. Governments must act as backstops or insurers to the project to satisfy the guarantees on completion and operation which lenders will require," Unruh stated. This need for government financial support, he continued, "should be communicated to both governments from the outset and on a continuing basis." He warned that "failure to commence dialog with the two governments involved very early will ultimately lead to significant delay of the project," and that, "it would be a serious error not to apprise our governments of this fact from the beginning."

At the time, other companies in the Arctic Gas group were still clinging to the hope that the pipeline could be financed without the insurance of government support. In the end, the financial advisors, major banks and investments firms in both countries were adamant in their opinion that the line could not be financed without government help and so testified before the regulatory authorities in both countries. The reliance on government financial help was one of the factors which ultimately led to the rejection of the Arctic Gas proposal and the approval of the alternative Alaska Highway pipeline. In 1979, the undecided fate of the Alaska Highway proposal still hung on government financial support.

When Sohio announced its withdrawal from Arctic Gas, the printing presses were already rolling, turning out the massive documents that would be filed with the applications. Some $50 million had by then been spent by the member companies, and the results of the studies were compiled in a five-foot stack of documents that would be filed in support of the applications. In Calgary, seven printing firms were busy turning out the documents that would be filed in Canada; in Washington, the U.S. printer had rented a three-storey warehouse to collate the material.

On the morning of March 17, 1974 a chartered Fairchild F-27 aircraft took off from Calgary, loaded with ninety-three hundred pounds of Arctic Gas filing materials. The flight itinerary called for deliveries of these materials to Toronto, Ottawa and Washington. At Washington, five thousand pounds of American materials were picked up, and the flight continued back to Ottawa, Toronto, Calgary, then on to Edmonton, Yellowknife, Whitehorse, Juneau and Anchorage. Federal, provincial, territorial and state governments were inundated with Arctic Gas materials, perhaps more than they cared to read.

The applications were filed in both Ottawa and Washington on March 21. In Ottawa, Canadian Arctic Gas Pipeline Limited had applied to the Department of Indian Affairs and Northern Development for a right-of-way across Crown lands in the Yukon and Northwest Territories and with the National Energy Board for a certificate of public convenience and necessity. In Washington, Alaskan Arctic Gas Pipeline Company had applied to the Department of the Interior for a right-of-way across U.S. federal lands in northern Alaska and to the American Federal Power Commission for a certificate of public convenience and necessity.

Voluminous though it was, the material supporting the Arctic Gas applications was still not complete. The supporting Arctic Gas exhibits related to engineering design, construction and operation; environmental matters; studies of alternative routes (including the Alaska Highway) and systems of transportation; and studies of the anticipated socio-economic effects in northern Canada and Alaska. Additional materials—filed during the following eight months as they were completed—related to cost, tariffs, gas supply and markets, financing, and anticipated impacts on the national economies of both countries.

On the day that the Arctic Gas applications were filed, the Federal Government issued an Order-in-Council appointing Mr. Justice Thomas R. Berger, of the Supreme Court of British Columbia, to conduct a public inquiry in relation to the application for the use of Crown lands in the territories for a pipeline right-of-way. Judge Berger was directed "to inquire into and report upon the terms and conditions that should be imposed in respect of any right-of-way that might be granted across Crown lands for the purposes of the proposed Mackenzie Valley pipeline." In formulating these recommendations, he was to have regard to:

(a) the social, environmental and economic impact regionally, of the construction, operation and subsequent abandonment of the proposed pipeline in the Yukon and Northwest Territories, and

(b) any proposals to meet the specific environmental and social concerns set out in the Expanded Guidelines for Northern Pipelines as tabled in the House of Commons on June 28, 1972.

The Order-in-Council asked Justice Berger "to report to the Minister of Indian Affairs and Northern Development with all reasonable dispatch." The first volume of Justice Berger's report was issued more than three years after his appointment, but the recommended terms and conditions contained in the second volume of his report required nearly four years from the time of his appointment. In 1977, two public inquiries on the environmental and socio-economic implications of the proposed Alaska Highway project took less than four months from the time the inquiries were appointed until the reports were produced.

Political reaction to the Arctic Gas applications was mixed. The Government of Ontario, concerned about the adequacy of gas supplies from Alberta, was strongly supportive. "Already in Ontario our gas utilities are constrained in their ability to enter into new contracts to supply natural gas for new users for significant expansions," Ontario Energy Minister Darcy McKeough told a meeting of corporate planners in Toronto early in 1974. "The hard fact is that the availability of natural gas can become a severe constraint in our ability to expand the industrial base of this province."

The New Democratic Party adopted its position early and stuck to it. Three months after the applications were filed, during the 1974 federal election, NDP leader David Lewis attacked the Arctic Gas proposal in a Yellowknife speech. Immediate construction of the line, he said, would be "an unjustified sell-out of Canadian resources." It would involve "terrible ecological risks" and would pose the greatest threat to native people "since the white man bulldozed his way into this territory." Never missing a chance to hammer at the detested multinational oil companies, he accused them of being "hell bent on exporting Canada's future gas supply to the United States.... The fact that Canada will not need that gas until 1990, but then will need it badly, is of no concern to these multinationals."

While Arctic Gas was the first horse at the starting gate, El Paso Natural Gas chairman Howard Boyd was already grooming another contender. In a speech to the Anchorage Chamber of Commerce, on December 4, 1972, Boyd announced that El Paso had begun studies of the feasibility of moving Prudhoe Bay gas eight hundred miles by pipeline to a port in southern Alaska where it would be liquefied for shipment by tanker ships. Such a system, he said, would provide Alaska with maximum benefits in the form of income, employment and tax revenues. Moreover, this "all-American" project would provide greater economic benefits to Americans than an overland pipeline which would require shippers of Alaskan gas to contribute substantial tax payments to Canadian authorities.

Less than two years later, on September 24, 1974, Boyd held a press conference in Detroit, where he was participating in a World Energy Conference, to announce that El Paso had filed an application for its Alaskan pipeline and tanker project with the Federal Power Commission.

The bedrock of El Paso's case was that the United States should not entrust Canada with the movement of its Alaskan gas. Boyd hammered this message out in his Detroit announcement: "The competing project suggests a 5,400-mile pipeline largely under for-

68

eign control," and which "would cut through virgin territory in Canada and a wildlife refuge in Alaska." This "would result in billions of dollars of tax and profit payments to Canada by U.S. consumers. Selection of a Canadian route would commit the future gas discoveries in the Arctic region of Alaska to dependence on a foreign sovereign.... The commitment of the great volumes of Alaska's North Slope gas to a pipeline controlled by a foreign power and traversing its heartland, no matter how friendly that power may appear, would repudiate the entire concept of U.S. energy independence."

The Federal Power Commission set May 5, 1975, for the start of consolidated hearings on both the Arctic Gas and El Paso applications, appointing Administrative Law Judge Nahum Litt to hear the evidence.

A third horse, being groomed by Bob Blair and Kelly Gibson, was brought to the starting gate on March 27, 1975, when Foothills Pipe Lines Ltd. filed its applications with the National Energy Board and the Department of Indian Affairs and Northern Development. Foothills applied for an eight hundred and seventeen mile, forty-two-inch diameter pipeline from the Delta gas fields to the northern Alberta border where the gas would be transferred to pipelines proposed by Alberta Gas Trunk Line and Westcoast Transmission. Foothills also proposed four hundred and sixteen miles of small diameter pipelines to deliver gas to northern communities, the cost of which would be subsidized by a surcharge on the transportation tariffs paid by consumers of Delta gas in southern Canada.

"The Foothills applications constitute the main part of the 'Maple Leaf Project' which its sponsors believe could provide an economical and the surest and most manageable means of connecting Arctic gas to markets served by the existing Canadian gas pipeline systems," the company's announcement stated. The Foothills announcement also said that, "some of the material filed as part of, or in support of this application will be identical in some cases and similar in others to material filed" with the Energy Board by Arctic Gas. It pointed to the information developed by Arctic Gas, up to the time that Trunk Line withdrew from the group.

As the Federal Power Commission consolidated the applications of Arctic Gas and El Paso for the hearings in Ottawa, the Energy Board set July 8th and 9th for a pre-hearing conference of applicants and all interested parties, with the actual hearings to start on October 27, seven months after the start of the Berger hearings and nearly six months after the start of the FPC hearings.

It was now a three-way race. In the United States, Arctic Gas faced competition from El Paso's "all-American" project; and in Canada, it faced competition from the "all-Canadian" project of Foothills. In place of building two separate transportation systems, one for American and a second for Canadian gas, Arctic Gas argued that the national interest of each country would be best served by co-operating to build only one line serving both countries.

By the time of the pre-hearing conference in Ottawa, seventy-five parties, in addition to the applicants, had registered for status as "interested persons," thus becoming participants in the hearings. Included were special interest and environmental organizations, provincial governments and corporate interests. Thirty-three of the intervenors said they supported early construction of a Mackenzie Valley gas pipeline, of which fifteen specifically supported the Arctic Gas proposal and one, Inland Natural Gas of British Columbia, tentatively supported the Foothills proposal. Twelve of the intervenors said they were either opposed to construction of a pipeline, or had some concerns or reservations.

The Energy Board had decided, on April 1, that Marshall Crowe would, in fact, chair the panel of three board members that would hear the applications. But at the time of the pre-hearing conference, this had not been announced. It had been eighteen months since Arctic Gas has apprised Crowe of the initial opinion it had received that there might be a legal problem if he decided to chair the hearings. Yet, there had been no statement as to whether Crowe would, in fact, participate in the hearings. Arctic Gas had decided it was time to pursue the matter further.

After the close of the second day of the pre-hearing conference, D.M.M. Goldie, QC, counsel for Arctic Gas, met with Hyman Soloway, QC, counsel for the Energy Board. It was very much a meeting of distinguished counsellors. Mike Goldie—rusty haired, tall, lean, athletic, immaculate in a three-piece business suit adorned with a gold watch chain—was a member of Vancouver's largest law firm, Russell and DuMoulin. Hy Soloway—more unassuming in attire, a stockier man with a round face and eyebrows that might have been stolen from John L. Lewis—was the senior partner in Ottawa's largest law firm, Soloway, Wright, Houston, Greenberg,

O'Grady and Morin. Staff lawyers had previously acted as board counsel at hearings, but this one promised to be something more than run-of-the mill, and so the board had retained one of the luminaries of the legal field. In addition to being one of Ottawa's top lawyers, Hy Soloway was also one of the pillars of the community: principal shareholder in Skyline Cablevision, vice-chairman of the board of Governors of Carleton University, a member of the advisory board of Guarantee Trust Co. of Canada, an executive member of the Canadian Jewish Congress.

This meeting of board counsel drew a response, six weeks later, in the form of a letter from board secretary Robert Stead to Goldie. Copies of the letter were sent to all those who had registered as participants in the hearings and made available to the news media.

Stead's letter recounted the July 9 meeting which had been requested by Goldie. "You indicated that, on instructions from your client, it was your intention to seek a ruling to determine the propriety of the composition of a panel of the board to deal with your client's application if the Chairman of the National Energy Board, Mr. M.A. Crowe, were a member," Stead wrote. "At no time was there any suggestion that there was, nor did you indicate that you feared, actual bias or proprietory or pecuniary interest insofar as the chairman was concerned. Your fears were directed to the possibility that, based upon the above-stated, any certificate issued to your client might be subject to attack on the grounds that there was a reasonable apprehension of bias by the chairman in favour of your client."

Stead's letter reported that Crowe had, indeed, been designated chairman of the hearing panel. However, "in order to allay any fears or reservations" it was decided that Crowe would "at the opening of the hearings make a statement to all participants and interested parties" outlining his prior association with Arctic Gas "and will hear objections, if any." The letter requested that Arctic Gas make available to the board and all interested parties copies of minutes of Arctic Gas committee meetings and correspondence with the Canada Development Corporation at the time Crowe was involved with Arctic Gas.

The procedure outlined by the board promised to eliminate the risk that a decision could later be attacked on this issue in court. If all the parties involved were apprised of the facts at the outset and waived any objections, it would be doubtful that the courts would then allow them to attack the decision later. The question was whether all parties would waive any objections. Arctic Gas, concerned that the board's hearings were already so far behind the other hearings, hoped that they would.

Therefore, in his response to Stead's letter, Goldie endorsed the board's approach to the problem: "The assurance that what the Board now proposes will not result in an attack after the decision rests upon the belief that our Courts would not permit such a proceeding to succeed when the facts had been discovered and accepted prior to the hearings commencement. While the law is not entirely clear on this point, I believe, however, that what the Board proposes would suffice and I repeat my appreciation of its decision to deal with this matter now." That is, if there were to be a legal attack, the board's procedure would flush it out at the start, rather than a year or two later at the time of the decision. If it had to come at all, better it should be sooner than later.

Newspaper reports at the time suggested that Arctic Gas had an entirely different motivation in raising the Crowe issue: "Sources familiar with the thinking of Arctic Gas management... wondered whether the motivation for the bias challenge might be the opposite —that Arctic Gas fears Mr. Crowe has nationalist leanings and would be hostile to their joint Canadian-U.S. project," Ian Rodgers wrote in the *Toronto Globe and Mail*. Another *Globe and Mail* writer, business columnist Ronald Anderson, concluded that "Arctic Gas has set out to remove Marshall Crowe" as chairman of the hearing panel. "It is difficult to believe that anyone could seriously fear that Mr. Crowe would be unduly biased in favour of Arctic Gas," Anderson wrote. "In fact, most observers suspect that Arctic Gas is afraid that the NEB chairman may be biased against its application, for nationalistic reasons."

Anderson also saw the issue as a problem for the board. "The NEB can hardly avoid having its chairman sit on the panel in a matter as important as the Mackenzie River pipeline," he wrote. "The panel, in such a situation, becomes the NEB, and its findings are binding on the National Energy Board. As chairman of the NEB, Mr. Crowe can hardly be expected to feel comfortable if a panel of which he was not a member made binding decisions of great import. He, after all, bears a considerable weight of responsibility for the NEB decisions."

A pall of uncertainty hung in the air as Marshall Crowe opened the board's Mackenzie Valley pipeline hearings beneath the crystal chandeliers in the faded grandeur of Ottawa's Chateau Laurier on October 27, 1975. More than two hundred people, including participants, spectators and newsmen, were on hand for the event. Cameras and tape recorders are normally prohibited in NEB hearings, the quasi-judicial proceedings usually conducted in the atmosphere of a court room. But to record this auspicious event for posterity, the television lights glared and cameras whirred, at least briefly, while Crowe, flanked at the presiding table by the other two panel members, Bill Scotland and Jacques Farmer, read a brief introductory statement:

The hearing itself is unprecedented in terms of its magnitude and pervasiveness of its issues. For the first time the Board is hearing applications to move gas from frontier areas. The filings reflect a wider scope of inquiry than any the board has undertaken. The wide spectrum of interest is represented, and in the course of making a decision on the applicants many related findings will have to be made, each of which will affect Canada's energy future.

Then the television lights were turned off, the cameramen carted their gear away, and Crowe got down to the issue of whether or not he should be there at all.

Crowe noted that all interested persons had been provided with copies of the exchange of correspondence between the board and Goldie, "relating to my participation at certain meetings of the Gas Arctic Northwest Project Study Group." He noted also that he had left the civil service in November, 1971, on appointment as a provisional director of the Canada Development Corporation, then wholly-owned by the government, and that the CDC had joined the Arctic Gas group in November, 1972. Crowe recited at length the details of his participation as the CDC representative at Arctic Gas meetings, noting that he had moved the resolution appointing the project's banking advisors; had participated in selection of the financial and accounting advisors; and had voted in favour of the crucial recommendation which later led to the withdrawal of Trunk Line from the group, the decision that Arctic Gas would build its own system across Alberta.

Crowe stated that, "my participation in the Study Group was at all times as a representative of this Government-owned organization and was but one of my responsibilities as president and chairman of the CDC." He concluded: "If, on the facts which I have stated and a review of the material to which I have referred, any person has a reasonable apprehension of bias on my part, that person has a right to object."

Assistant board counsel, Ian Blue, announced that each participant would be asked to "state on the record whether he has or has not any objection so far as the chairman's position is concerned." Blue then called the roll, starting with Arctic Gas.

Blue: "Does Canadian Arctic Gas Pipeline Limited have any objection to Mr. Crowe sitting on the panel?"

Goldie: "No."

One by one, more than fifty participants at the hearing all indicated that they had no objection to Crowe's chairing the hearing panel. It was starting to look as though there would be no legal problem after all. Then Anthony Lucas responded for the Canadian Arctic Resources Committee: "Mr. Chairman, I have been instructed to formally object."

In addition to expressing concern about possible apprehension of bias on the statements on record, Lucas had "one additional matter" to raise. This was based on a forthcoming book, *The National Interest*, by Professor Edmund Dosman of York University, portions of which had been printed in a series of articles in the *Toronto Star*. One of the articles, said Lucas, referred to a meeting on May 12, 1970, "involving a number of senior federal public servants, including Mr. Crowe, then in his capacity as a senior official of the Privy Council Office ... That meeting, Professor Dosman suggests, was critical in hammering out the essential content of the 1970 Northern Pipeline Guidelines" which were claimed to "amount to approval in principle for a Mackenzie Valley gas pipeline." According to Lucas, this provided further grounds for apprehension of bias.

John Olthuis responded to Blue that the Committee for Justice and Liberty was not concerned about "a reasonable apprehension that Mr. Crowe may favour the application of Canadian Arctic Gas Pipeline Limited over the application of Foothills Pipe Lines." The CJL was concerned, however, about "a reasonable apprehension that Mr. Crowe may be biased" in favour of the need to build any pipeline. An energy work group from York University adopted a similar position, and the Consumers Association of Canada suggested that the matter should be referred to the Federal Court for a ruling.

74

The first day of hearings was adjourned until nine the next morning, while the board figured out what to do in the face of this challenge. They had not yet figured it out when the hearing reassembled the next morning. Soloway read a brief statement: "At the conclusion of yesterday's session we indicated that there would be an adjournment until nine o'clock this morning. Because of certain issues that were raised yesterday and the fact that we are still in the process of checking certain facts that were put on the record, I have been instructed to advise you that the board will resume tomorrow at nine o'clock." That was it. The second day of hearings lasted less than four minutes.

On the third day, Crowe read another opening statement: "The board considers it of the utmost importance that the issues giving rise to the objections be determined through proper judicial procedures, expeditiously, and in a manner consistent with confidence in the administrative law processes of Canada ... the issue as to whether there is a reasonable apprehension of bias in the panel appointed by the board, of which I am a member and its chairman, will be referred ... to the Federal Court of Appeal for hearing and determination. I will be guided by the decision of that court." Meanwhile, the hearings were to continue.

Once again, news accounts darkly hinted that Arctic Gas was out to get Marshall Crowe. *Globe and Mail* correspondent Jeff Carruthers suggested that the special interest groups were the victims of circumstance. On a CBC radio programme, Carruthers said:

> The consortium really fears Crowe's nationalist leanings and that he may side with Foothills, the proponent of a cheaper all-Canadian pipeline. What's disturbing is that now the public interest groups are falling into the inexcusable trap of doing the dirty work for Canadian Arctic Gas, which would obviously benefit from Crowe's removal. No one has said yet that Crowe is biased, only that some people might think there was a possibility of bias. Yet if he is nationalistic, as Canadian Arctic Gas seems to fear, then that would seem to be more reason for him to stay than to leave. The National Energy Board now more than ever needs to assert itself as acting in the Canadian national interest, especially if it is to make up for the many mistakes it had made when U.S. interests dominated, producing the energy shortage problems Canada is now facing.

The question put to the Federal Court by the board was: "Would the Board err in rejecting the objections and in holding that Mr. Crowe was not disqualified from being a member of the panel on the groups of reasonable apprehension or reasonable likelihood of bias?" The Federal Court found in favor of Crowe and the board, and the hearings continued, until the matter was pursued further in an appeal to the Supreme Court.

It was the religiously-based Committee for Justice and Liberty that appealed the Federal Court's ruling to the Supreme Court, supported by a loosely-organized number of special interest groups banded together as the Public Interest Coalition or PIC. PIC had come into being as a result of the efforts of various organizations to secure public funding to finance their participation in the board's hearings. During the Berger Inquiry, native, environmental and other organizations had secured nearly $2 million in government funding, and the same type of assistance was being sought to attend the NEB hearings. PIC members included the Committee for Justice and Liberty, the Canadian Wildlife Federation, the Committee for an Independent Canada, Energy Probe, the Consumers' Association of Canada, Canadian Arctic Resources Committee, York University's Workshop on Energy, and others.

Some time after the hearings, I spoke to Professor Ian McDougall of Osgoode Hall Law School, York University. He had acted as counsel for the Canadian Wildlife Federation and as chairman of PIC at the board hearings. "We couldn't get the Berger experience duplicated," McDougall told me. "We tried. In fact, we tried every trick under the sun to do it. We felt that it wasn't doing anyone any service going in there unprepared. The only way to be prepared would be to have professional people," working at it full-time. "We wanted to do a good job, and we weren't going to be able to do it because we weren't going to be given any help. It was just unrealistic to assume that we could get adequate funding from our organizations . . . In some cases we tried and just failed horribly. It looked as if we had no choice to co-operate."

At the outset, the Committee for Justice and Liberty was the only organization in PIC determined to appeal the Federal Court ruling, according to McDougall. Most of the other organizations supported Crowe, he said, in part because they felt Crowe might support their search for funds. McDougall also said that several representations were made to Crowe requesting funding, but "he didn't give an inch."

"It was suggested to a couple of us... at the first PIC meeting that we would be well advised if we stated the case in Crowe's favour," McDougall recalled. "It was raised in response to that suggestion that there was no money to do that sort of stuff. The counter response to that was, 'oh, well, maybe money could be made available for a project which was that meritorious. That got a lot of people mad. That story got repeated. In fact, I made sure it got repeated to everybody involved. There never was any evidence that Crowe was directly involved in this offer." Nevertheless, the PIC organizations, in McDougall's view, felt that they were being "screwed," and thus unanimously supported the appeal brought before the Supreme Court by the Committee for Justice and Liberty.

If the Supreme Court ruled in favour of CJL, it would mean that the board's hearings would have to start all over again, under a new hearing panel. A lot of time would be lost. With the FPC racing ahead with its hearings in Washington, and El Paso holding out the promise of early delivery of Alaskan Gas, time was crucial to the Arctic Gas case. In the three days of hearings before the Supreme Court, Goldie argued in support of Crowe and the NEB. So did counsel for Foothills, Trunk Line, TransCanada PipeLines, Alberta Natural Gas Co., and the Attorney General of Canada. It was all to no avail.

On April 12, 1976, the new Energy Board hearing started with a panel of three board members, J. G. Stabback as hearing chairman, C. G. Edge, and R. F. Brooks. The board's hearings were now more than a year behind Berger in Yellowknife and the FPC in Washington.

Chapter 6
The Panama Canal of Canada

Yellowknife, March 3, 1975. The Explorer Hotel, owned by the Government of Alberta, sits on a small rise of gleaming, snow covered Precambrian rock, overlooking a vista of more rock with stunted trees and the City of Yellowknife. Like Yellowknife itself, the hotel is a modern urban anomaly in the endless expanse of northern wilderness. The large ballroom, where the Legislative Council of the Northwest Territories sits when it is in session, is packed solid with hundreds of people from the north and from across Canada. Included are scores of newsmen and television crews, more than have ever been in Yellowknife at one time before. After a year of preliminary hearings on procedural matters, organization, and northern familiarization visits, Mr. Justice Thomas R. Berger, of the Supreme Court of British Columbia, is about to open the first day of formal hearings of the Mackenzie Valley Pipeline Inquiry.

It is also the day that Foothills Pipe Lines Limited, led by Alberta Gas Trunk Line and Bob Blair, will warn of the "threat and danger to the whole of the Canadian people and the Canadian nation" posed by the prospect of pipelining American Alaskan gas across Canada. For four years, Blair had actively promoted just such a pipeline—"the future main North American artery for natural gas transport," he had called it. Now he maintained it had become clear that such an undertaking would pose serious risks for Canada, so serious that such a pipeline would be "the Panama Canal of Canada." Thus, Foothills was seeking approval for a smaller pipeline which would transport only Canadian gas from the Mackenzie Delta to Canadian consumers, with perhaps just a little bit for export sales to the United States.

March 3rd was a day of opening statements by counsel for the inquiry applicants, Arctic Gas and Foothills, and by counsel for the other hearing participants, the native, environmental and other northern organizations. They were statements that set out the issues in broad, often dramatic terms, perhaps intended as much for the benefit of the national television audience as the edification of the judge who sat impassively at the table at the front of the room. But first, it was Judge Berger's turn at bat. "We are embarked on a consideration of the future of a great river valley and its people," the judge said. "I want the people who make this valley their home to have a chance to tell me what they would say to the Government of Canada if they could tell them what was in their minds. Then it will be my task to report to the Minister of Indian Affairs and Northern Development regarding the social, economic and environmental impact of the project here in the north and to recommend the terms and conditions that should be imposed with respect to any right-of-way that may be granted for the pipeline. It will be for the Minister and his colleagues—for those who govern our country—ultimately to determine whether there should be a pipeline, and the terms and conditions that should be imposed with respect to any right-of-way."

First to present a statement was Pierre Genest, QC, who spoke for Arctic Gas. A leading Toronto lawyer and former appointed member of the NWT Legislative Council, Genest brought to the hearings Churchillian oratory and a neat turn of phrase, leavened with a touch of appealing humour. (He was also a man of prodigious proportions. Absent from the hearings for a period of time, he returned having lost considerable weight. "I left as Falstaff," he explained to the hearing, "and return as Cassius".) On this first day, he was accompanied by three other lawyers: John Steeves of Vancouver, Jack Marshall of Calgary and Darryl Carter of Yellowknife. It might seem, said Genest, like "an excessively large gaggle of lawyers," but with three public hearings on at the same time, it might be necessary for some of the lawyers to be detached from time to time for the NEB and FPC hearings. Mike Goldie, who had appeared at the preliminary hearings, was absent, having, Genest explained, "disappeared into the National Energy Board."

"The Arctic Gas companies were brought into existence by reason of the concern of its sponsors of a possible shortage of natural gas," Genest said. "That possibility has now become a fact ... An amount in excess of seventy-five million dollars has now been spent by the sponsor groups in order to carry out what I can call without hesitation the most extensive and thorough preliminary engineering, environmental and sociological study ever given to any project anywhere." Genest outlined the case which Arctic Gas would present and which, he said, would demonstrate that the pipeline was required in the national interest and would be of great benefit to northern residents.

Next up was Reginald Gibbs, QC, Calgary, counsel for Foothills, who argued his case with what he described as his "country logic," because "I don't have the Toronto sophistication." He may have seemed more like a country school teacher, with his rimless glasses and ample moustache, but Reg Gibbs was a former lawyer with the Canadian subsidiary of Standard Oil Company of California; he brought to the hearings energy and courtroom skill that few could match.

"The Canadian Arctic Gas project is not designed to meet Canadian needs," but to transport northern gas to U.S. markets, Gibbs argued. The United States would "come to rely" on such a delivery system. "There will be schools and hospitals and residences and businesses who depend on what is carried through that new highly visible four-foot tunnel from the Arctic, and that, sir, will rapidly assume the character of a very visible and indispensable lifeline."

A north-south transportation artery like that proposed by Arctic Gas "runs directly counter to Canadian historical experience," Gibbs argued. "Throughout our history as a nation there has been insistence that main Canadian transportation and communications systems follow all-Canadian routes, with only convenience connections across the United States Border." The only exception to this, Gibbs claimed, was the Columbia River Treaty, of which "the consequences are spoken of in sorrow... This insistence on all-Canadian routes... has been one of the greatest single factors which has preserved the integrity and sovereignty of Canada. It seems incredible to me that there would now be proposed at this stage in our history that we should abandon all we have learned and favour construction of what has been described as the equivalent of a Panama Canal across western and northern Canada." Canada will need gas from the Delta "at the latest by 1979 and probably a year or two before," and ought to "restrict the use of Crown lands to wholly-Canadian use" with "the carriage of Canadian gas to Canadian markets through Canadian systems," Gibbs continued. He concluded with an urgent plea: "With all the force that I command I urge, sir, and plea that we avoid the dedication of a north-south corridor, no matter how narrow, to foreign use and control in perpetuity."

Five months later, on August 18th, Blair appeared before the Berger Inquiry with a slightly modified version of Foothills' opposition to moving Alaskan gas. It might, after all, be possible to move just a part of Alaskan gas production through the proposed Maple Leaf pipeline: "Personally I can see it possible to extend the use of that system under some circumstances to provide some service for the movement of Alaskan gas across Canada, if the Government of Canada should decide that they wish that offer to be made to the United States."

But the task of moving all the gas that might be produced on the Alaskan North Slope still presented some horrors for Canada. "I would hope that the Canadian utilities would not be asked to move all of the gas which would be produced in Alaska," Blair said. "I have come to the view that such an undertaking would, in both the immediate future and the long run, bring more problems and hazards into Canada than it would bring in advantages of commercial business for the pipeline companies." This would include "the enormous and continuing responsibility of raising capital not to serve Canadian markets but strictly to enlarge a system to serve the purposes of U.S. companies, particularly in times when projects of much greater significance or need to Canadian purposes could be urgent." Another danger would be "the responsibility of steering each installation and expansion through regulatory and rate-making processes and procedures and sovereign concerns of jurisdictions which are properly autonomous and which could have different preoccupations from time to time." It would be far better to build an all-Canadian line for the movement of Delta gas so that "the timing of much needed Canadian supply to Canadian markets will not in any way depend upon what happens in regulatory litigation or legislative processes elsewhere."

The following day, August 19th, 1975 Blair was questioned further by Glen Bell, Counsel for the NWT Indian Brotherhood, on the possibility of moving Alaskan gas through the Maple Leaf line. "We do not want that job," Blair responded. "We don't seek it, we didn't apply for it, we wouldn't want to move all the Alaskan gas anyway. We are negative about moving all the Alaskan gas because we think that...produces more problems and hazards that it does any business advantage."

Bell: "If you did move some Alaskan gas, how would you envisage the rest of it getting to the United States?"

Blair: "Well, we wouldn't care. I mean it wouldn't be our business."

Commission counsel Ian Scott asked: "I take it from what you have said there is no circumstance that you can envisage in the next decade in which you would carry the volume of Alaska gas that the Arctic Gas consortium proposes to carry?"

Blair: "There is certainly no circumstance that I can perceive of any probability that we would want to do that job."

As late as February, 1976, Blair was still expressing opposition to building a pipeline to move Alaskan gas across Canada. He told the Federal Power Commission that he was uncertain whether it would be possible to raise $700 million equity in Canada as proposed by Arctic Gas. "The real concern, however," he said, "should be...the willingness of Canadians to subscribe to that amount of equity for a project which is manifestly for the primary benefit of the United States economy."

At the same time, Westcoast president Ed Phillips asserted that, "The capital requirements of Canadian Arctic Gas would create an undue burden on the Canadian capital market, thereby making it difficult, if not impossible, for Canadians to finance other worthy projects which might be in their interest." Yet the time was fast approaching when Blair would again advocate the undertaking which he had told Judge Berger he was so adamantly opposed to.

On May 5, 1976, Foothills, Alberta Gas Trunk, Westcoast and Northwest Pipeline signed a letter of agreement to seek approval for a new $7.3 billion pipeline to transport Prudhoe Bay gas across Canada to U.S. markets by a route following the Alaska Highway. The proposed forty-two-inch diameter line would total two thousand seven hundred miles, including seven hundred and thirty-one miles across Alaska to be built by Northwest, five hundred and thirteen miles across the Yukon to be built by Foothills, and fifteen hundred miles across British Columbia, Alberta and the southwest corner of Saskatchewan to be built by Westcoast and Trunk Line.

Bob Blair's northern pipeline proposals had now come nearly full circle, from the first proposal for a line to move both Prudhoe Bay and Delta gas, to a proposed line to move just Delta gas, and now another proposed line designed to move just Prudhoe gas. A principal difference between the first and third proposals was that the latter route was seven hundred and fifty miles from the Delta, thus requiring two pipelines to move the gas from both supply areas. The agreement of May 5th, in fact, stressed that the "first priority" was to be given to the Maple Leaf line to move Delta gas. This first priority would apply even though it was intended that the Maple Leaf line would be constructed after the Alaska Highway line.

The "Panama Canal" proposal of Arctic Gas, two hundred and seventy miles shorter than the Alaska Highway route, would move gas from both supply sources. The Alaska Highway route could move only Alaskan gas, unless it were connected with a lateral spur from Whitehorse along the Dempster Highway to the Delta. Such a lateral would in itself be no slight undertaking. At seven hundred and sixty miles it would be virtually as long as the Alyeska oil pipeline, which cost $9 billion to build, albeit for a larger diameter line.

In a statement commenting on the Alaska Highway pipeline proposal, Bill Wilder said it would make about as much sense to build two pipelines—one for American gas and one for Canadian gas—as it would to build two St. Lawrence Seaways, one for American ships and a second for Canadian ships.

The trail leading from a position of not wanting to undertake the transportation of all the Alaskan gas to an agreement advocating exactly that, was blazed with further Blair assertions. On August 7, 1975, the *Edmonton Journal* quoted Blair as stating that Foothills would consider selling controlling interest in its Maple Leaf line to native organizations. He suggested the estimated cost of $160 million could be secured from the settlement of native land claims. On February 13th, he told a seminar at the University of Calgary that Foothills was prepared to pay several times the normal fees for land uses for right-of-way, providing hundreds of millions of dollars which could form part of a negotiated settlement of land claims. This would allow native organizations to acquire large blocks of Foothills shares, permanent representation on its Board of Directors, and some control over hiring. On December 10th, Blair told the National Energy Board that Foothills would not build any pipeline in a region where as few as two per cent of the "responsible" residents were opposed:

Where even a significant minority, even a five percent, two

percent of the responsible local residential citizenry says 'We don't care what permits you have got, we don't care what licence or what legal documents you are holding in your hand, we do not want you in here, and get off.' We don't go in there. We talk and keep on talking and persuading until we can get in, and that is a truth of responsible pipeline management.

The Alaska Highway route was one of a dozen alternative routes examined in the studies filed by Arctic Gas with the regulatory authorities when the applications had been filed in March, 1974. It had been rejected as too costly, if both Prudhoe Bay and Mackenzie Delta gas were to be brought to market. As an acceptable political option, however, it seemed increasingly attractive in late 1975 and 1976. It was the route favoured by American environmental organizations, which were adamantly opposed to crossing the Alaska Wildlife Range, which the Arctic Gas route involved. And the route past Fairbanks and several smaller Alaskan communities, provided at least some measure of the benefits which the State of Alaska saw in the El Paso proposal, which it favoured.

Westcoast Transmission, through its participation in the Mountain Pacific group, had looked at the Alaska Highway route. In evidence filed with the Federal Power Commission in February, 1976, Westcoast president E. C. Phillips reported that, "The Mountain Pacific group has not been disbanded. In fact, it has recently taken a preliminary look at a route for a pipeline to carry only United States gas from Prudhoe Bay through Fairbanks, southeastward into the Yukon Territory through British Columbia. At the present time, however, it is not pursuing this project in an active manner."

Phillips, in fact, had written to Alaska Senator Ted Stevens on October 14, 1975, pointing out the claimed advantages of the Alaska Highway route: "From the time we started to study our initial plan to move Alaskan gas, called Mountain Pacific, I have felt very deeply that the United States should have the security of its own pipeline to the lower 48 states carrying exclusively Alaska gas." Phillips said he was convinced that an Alaska Highway route "could deliver gas to a common point on the U.S. border at a lower unit cost" than the Arctic Gas route "although it is a longer pipeline." A brief tabulation in the letter to Stevens showed the estimated transportation cost for the Alaska Highway route to be six per cent less than for the Arctic Gas route. It also showed a volume of four billion cubic feet a day of Alaskan gas being moved through an Alaska Highway line and only two billion through the Arctic Gas line. How the Alaska Highway line would be able to secure twice the volume of gas supply was not explained. "This cost advantage is something that should be explored in great depth," Phillips continued, "but I suggest the alternative route would be desirable from your standpoint even if the transportation costs were higher."

The Federal Power Commission staff produced a flurry of position papers, some of which supported the Alaska Highway route on environmental grounds, although the final FPC staff position called the Arctic Gas proposal "vastly superior." A draft environmental statement by the FPC staff in December, 1975, suggested that Arctic Gas should change its route to follow the Alaska Highway. Three months later, five environmental organizations, the Sierra Club, the Wilderness Society, Friends of the Earth, National Audubon Society, Defenders of Wildlife, and the Alaska Center for the Environment, petitioned U.S. President Ford by telegram to block the Arctic Gas line across the Alaska Wildlife Range and consider instead the Alaska Highway route. They also wanted the boundaries of the Wildlife Range extended and the area withdrawn from any application of mineral leasing laws.

While environmental organizations pushed for the Alaska Highway proposal and the State of Alaska fought for the El Paso proposal, Congressmen from the eastern and midwest United States lined up in support of the Arctic Gas project. Earlier they had sought to have the oil pipeline laid across Canada and Alaskan oil delivered directly to their region. Now that the Alyeska line was being built to move the oil to the west coast, they were determined not to lose out on a gas pipeline. Walter Mondale, the Democratic Senator from Minnesota who would become the American Vice-President before the year ended, introduced legislation to have Congress mandate the Arctic Gas project. Mondale's bill would take the decision out of the hands of regulatory authorities and vest it in Congress.

Mondale told the Senate that the El Paso plan "would rely on a vast system of displacement that is yet to be shown legally possible or technically feasible, except at great cost." He accused El Paso of raising "the bogus issue of Canada's reliability." Canada, he said, would make its own decision, "and in no way are we attempting to pre-judge what the Canadians will do." But "if Ottawa favours a joint undertaking—something that would help relations between the two countries which haven't always been smooth recently—we want to be in a position to move."

"The bill is a monster," responded Senator Ted Stevens, the indefatigable scrapper for Alaskan interests and Congressional champion of the El Paso project. "I think they have got the bear by the tail with this legislation. If the environmentalists don't rise up in righteous indignation because of this bill, I'll be very surprised."

By mid-February, twenty-eight Senators had lined up in support of Mondale's bill, and only one, Senator Clifford Hansen of Wyoming, had supported a bill introduced by Stevens to mandate the El Paso project. A third bill, introduced by the administration, proposed that the decision should be left to the President, subject to ratification by Congress, with a deadline for selecting a route to move Alaskan gas.

As Arctic Gas and El Paso lobbied Congress for support of their respective projects, Westcoast, Trunk Line and Northwest Pipeline shaped their Alaska Highway proposal. It was Trunk Line and Westcoast Transmission who first approached Northwest Pipeline with the idea of an Alaska Highway project, Northwest president John McMillian testified before the FPC. The two companies met to discuss this possibility on March 15, 1976. Six weeks later, on April 30th, Westcoast president Edwin Phillips wrote to Northwest concerning what was then described as Northwest's proposal to move Alaskan gas via the Alaska Highway. If Westcoast and Trunk Line were to participate in such an undertaking, Phillips wrote, "it would have to be established without any doubt or equivocation . . . that the group's first priority is its Maple Leaf Project" to transport Delta gas. Given this understanding, "AGTL and Westcoast are pleased to be identified as advocates of the Fairbanks Corridor— Alaska Highway alternative," Phillips declared. This proposal, he said, "will help to remove any unfortunate conception in the United States that those two companies' aggressive pursuit of the Maple Leaf project has an entirely nationalistic thrust, is based on obstructing the movement of U.S. gas across Canada, and is insensitive to the unusual gas shortages of our neighbour. . . . AGTL and Westcoast maintain that a properly designed Fairbanks Corridor alternative utilizing existing systems is supportable." It would be "a new international project purged of the deficiences of the Arctic Gas international plan now before the NEB."

While the May 5th letter of agreement identified Trunk Line and Westcoast as "advocates" of an Alaska Highway line, the companies at the same time agreed to "commence all studies necessary to determine the feasibility of such a proposal." It was further agreed that, "Appropriate filings will be made with the Federal Power Commission and the National Energy Board upon a determination of the feasibility of the project."

By this time, the whole idea of an Alaska Highway route had already been dismissed in yet another of the unending stream of FPC staff reports. "We could not advocate or support a Fairbanks alternative," wrote FPC staff members David Lathom and James M. Keily, Jr., "Our studies led us to believe that it was unwise and unrealistic to mount a major effort to study the engineering and economic feasibility of this alternative and thereby commit a significant portion of the staff's resources on an alternative that showed minimal prospects of bearing fruit."

The Trunk Line-Westcoast-Northwest group concluded otherwise. Within nine weeks of their May 5th agreement, they had not only decided that such a system was feasible but had also assembled enough material to file an application with the Federal Power Commission. Judge Litt ruled that the application by Northwest Pipeline would be consolidated in the hearings on the Arctic Gas and El Paso applications, which had already been in progress for fourteen months and were, in fact, nearing an end. The FPC estimated that this latest entry would add another sixty days to the regulatory hearings.

Although the application by Northwest was filed in Washington with the FPC on the 9th of July, it was nearly two months later, on August 31st that Foothills filed the corresponding application in Ottawa with the Energy Board. This delayed filing in Canada produced some heated arguments at the Energy Board hearings. Reg Gibbs filed a motion with the board requesting, among other things, a six-month extension of the deadline for filing certain cost estimates for the Maple Leaf project. Mike Goldie responded with another motion asking the board simply to dismiss Foothills' application for Maple Leaf.

Goldie accused Foothills of seeking to defeat the Arctic Gas project by delaying proceedings in Ottawa, while rushing forth with evidence in Washington. Where two years ago, Foothills had argued that Canada would need Delta gas probably as early as 1978, Foothills witnesses were now telling the FPC that, "Canadian markets can be served for the next six, seven or eight years out of gas presently being produced in Alberta." Because of this reassessment of the gas supply-demand picture, Foothills had told the FPC it would be desirable to build the Alaska Highway line first and the Maple Leaf line later. Goldie argued that, "... the determination of whether that is in the public interest in Canada should be done here. It should not be argued out before the Federal Power Commission." He also said that, "these people are chattering like magpies down in Washington and not a sound is heard up here."

"Mr. Blair, who heads the Foothills consortium which also proposes to build an all-Canadian Mackenzie River gas pipeline, says the belief is growing in U.S. government and industry that the Canadian government will not be able to make a decision on the Mackenzie Valley route in time for a joint Canadian-U.S. project to meet U.S. needs for gas from Alaska," Goldie asserted. "This would leave the United States with only two projects to consider, El Paso and Alcan [Alaska Highway route]... I cannot overlook the fact that Mr. Blair states that delay may result in the elimination of my client's proposal, not on the basis of its merits but on the basis of the regulatory process."

Later, Goldie observed that the Maple Leaf project was being buried, in favour of the Alaska Highway line, "without even the usual lamentations that accompany a funeral." He said, "the Maple Leaf project begins to resemble the Cheshire cat; it is slowly disappearing from sight, leaving behind it only Mr. Blair's grin."

The special interest organizations, however, saw the Arctic Gas arguments as an attempt to pressure the board into rushing through the hearings with undue haste and inadequate consideration. Professor Robert Page spoke for the Committee for an Independent Canada when he said that Arctic Gas had performed "a snow job on the country," by asserting "that Canada must speed up its deliberations in order that the Canadian decision will not be later than the American." He said it would be impossible for the United States to reach a decision "before the final weeks of 1977." The real reason behind the Arctic Gas statements, he claimed, was "because the faster the hearings the less the critics can find out, the less evidence anyone outside the two applicants can present, and the shorter will be the time span for consideration of the Alcan, the Polar Gas or any other alternatives." Francoise Bregha, speaking for the Canadian Wildlife Federation, asserted: "We have been denied funding. We have been denied access to government studies. Now we are denied time."

Through the summer months of 1976 the hearings droned on, before the Federal Power Commission in Washington, before the National Energy Board in Ottawa, and before Justice Berger in villages and cities from Tuktoyaktuk to Vancouver to Charlottetown. In the sticky humidity of Washington, committees of the Senate and House debated The Alaska Natural Gas Transportation Act of 1976. Congress was due to adjourn at midnight, Friday, September 30th, and would not be in session again until a new Congress assembled in 1977, following the Presidential and Congressional elections. The Act was adopted at 10:30 pm, just ninety minutes before the deadline, and sent to President Ford for his signature. The Act set a deadline for an American decision on a route to move Alaskan gas and established procedures to meet that deadline. Under the terms of the Act, the Federal Power Commission would not decide on the applications before it, but would instead provide a recommendation to the President. The deadline for that recommendation was set as May 1, 1977. Other government departments and agencies—State, Defense, Treasury, Justice, Interior, the Council on Environmental Quality—were to make their views known to the President by July 1st. The President was required to make his decision by September 1st, although there were provisions whereby he could extend this for ninety days. Congress would have sixty days to confirm or reject the Presidential decision. The Act contained limitations on any judicial review in order to avoid having the final decision tied up in the courts for an indeterminate period.

On November 12th, Judge Litt closed the record on the FPC proceedings after two hundred and fifty-three hearing days. Six days later, in Yellowknife, Justice Berger closed the record on the Mackenzie Valley Pipeline Inquiry after two hundred hearing days.

"The issues facing the Inquiry are profound ones," Judge Berger said in his closing remarks. "There is no consensus among northern people about these issues. There was no consensus when we began, there is no consensus today. We in Canada think of ourselves as a northern people. The future of the north is important to us all. What happens here, here on the northern frontier, here in the northern homeland, will tell us something about what kind of country Canada is, what kind of people we are. The Inquiry stands adjourned. You will be hearing from me."

Straining to catch up and produce a report on time, the National Energy Board extended its hearing hours, sitting from 8 am into the early evening. Lawyers and witnesses worked feverishly through the night, preparing for the next day's session. As the board's hearings continued into 1977, every indication seemed to be that Arctic Gas was riding the crest of a wave which would carry it through as the winner.

Even the State of Alaska, champion of the El Paso proposal, appeared to find that the Arctic Gas proposal offered the best economics. Late in 1976, Attorney General Avrum Gross, chairman of the state's Alaska Natural Gas Pipeline Task Force, delayed a planned report, when studies showed that Arctic Gas offered the lowest cost transportation. The studies, conducted by the state's Pipeline Surveillance Office, estimated the minimum transportation costs for Alaskan gas at $1.17 per thousand cubic feet for the Arctic Gas line, $1.74 for the El Paso system, and $1.65 for the Alaska Highway line. Gross observed that the lower transportation costs would increase the state's royalty revenues.

This was the time, December, 1976, that the FPC staff called the Arctic Gas route "vastly superior."

> The fundamental logic of the Arctic Gas route is unassailable. Even a casual view of a topographic map of North America will reveal that the sponsors of the Arctic Gas project have chosen the most logical natural gas pipeline route from Prudhoe Bay across western Canada to the central United States. When the existence of a natural gas field in the Canadian Mackenzie Delta is given recognition, the choice of the Arctic Gas route becomes overwhelming in its appeal.

The report also said that the forty-eight-inch diameter high pressure line proposed by Arctic Gas "is a consistent next step in the evolution of high-pressure natural gas pipeline design ... There will be problems, there are significant uncertainties; but the alternative is dramatically higher transmission costs. The risks are prudent and should be taken, in our opinion." The report further noted that the Alaska Highway and Maple Leaf lines would require an extra twelve hundred miles of pipeline to transport gas from both supply sources "in an enormously inefficient way."

Commenting on this report, Blair said that it "has completely missed the point on several items of great consequence to Canada ... Our creed is to get a manageable job identified and then get it done right, and these objectives are clearly best served by the Alaska Highway project in respect of gas from Alaska, as well as the Maple Leaf project for the Mackenzie Valley when the social and political environment there can make a pipeline conscionable."

El Paso vice president John Bennett said the FPC staff had behaved "as if Canada were sovereign American territory."

The first months of 1977 brought record cold temperatures to much of the United States; crippling shortages of natural gas; a flood of petitions to President Carter urging approval of the Arctic Gas project; signing of the Canada-United States pipeline treaty; and a four hundred and seventy page report from Judge Nahum Litt recommending approval of Arctic Gas and utterly rejecting the Alaska Highway proposal.

The combination of unusually severe weather and natural gas shortages closed thousands of schools, offices, shopping centres and factories throughout much of the United States. More than one million American workers were temporarily laid off. Canada responded with emergency deliveries of gas to the limit of its pipeline delivery systems.

It was an auspicious time for the signing of the Canada-U.S. pipeline transit agreement, initialed in Washington January 28, 1977, by Canadian Ambassador Jake Warren and U.S. Assistant Secretary of State Julius Katz. The agreement, first proposed by Prime Minister Trudeau in the House of Commons more than two years before, was designed to "confirm to both countries a regime of non-interference and non-discrimination for transit pipelines carrying oil and natural gas destined for one country across the territory of the other." The agreement was also seen as helping smooth the way for approval of the Arctic Gas project.

Three days later, Judge Litt tabled his report, based on the FPC hearing record of more than forty-five thousand pages plus more than one thousand exhibits (some of which in themselves exceeded one thousand pages).

Litt wrote, "there is a consensus on the part of commission staff, the most populous consuming states taking an active interest, and an array of pipelines and distributors serving huge sections of the country that if any pipeline applicant must be chosen now, their best interests would be served by choosing Arctic Gas. The evidence in this record clearly supports that conclusion...The Arctic Gas application is superior in almost every significant aspect when compared to El Paso." He added that El Paso, however, would be in the U.S. interest, "if it were not for the clearly superior Arctic Gas application. Thus, if Arctic Gas is unable to accept a certificate, this record supports findings that El Paso's proposal...would also meet the present and future public convenience and necessity."

"No finding from this record supports even the possibility that a grant of authority to Alcan can be made...Alcan's present design is clearly neither efficient nor economic since the pipeline is undersized. The suggested three years construction scheduled to be completed by 1981, which Alcan argues is one of its prime strengths, cannot occur. As presently proposed, even with Alcan's willingness to build anything anyone wants, as long as it does not oust Westcoast and AGTL from their Maple Leaf project, there is not enough left of its original proposal to serve as a basis for granting its application."

Litt noted further that, "the other applicants and staff criticize Alcan's engineering and geotechnical design for being so unsupported by meaningful design preparation as to make it extremely difficult, if not impossible, to determine the feasibility of its proposal. This criticism is painfully accurate." Litt concluded that summer construction of the Alaska Highway line "cannot be accomplished without unacceptable environmental impact." He was also concerned about laying a gas line close to the elevated Alyeska oil pipeline. Blasting a ditch within twenty-five to fifty feet of the vertical supports for the oil line would present some risks, because "...a mistake in Alcan's blasting could result in a break releasing substantial quantities of hot oil."

Litt also had some harsh words for the oil companies holding the gas reserves at Prudhoe Bay. He said that their support, as well as that of the State of Alaska, could be crucial to financing any line to move the gas, but "neither has shown any particular interest in such financing." In fact, "the producers have been downright hostile to the suggestion." Litt suggested that Congress might be able to find some "legislative methods" which would encourage the participation of the oil companies. "The corollary of not being able to make a horse drink when led to water is that you can make him darn sorry that he did not," Litt declared.

The judge also blasted the oil companies for failure to contract for the sale of their North Slope gas reserves, suggesting they were holding out for concessions from the FPC or Congress, and accusing them of placing their corporate interests ahead of the national interest: "...the producers, like G.B. Shaw's dinner companion, have a price at which they would sell their 'service,' and all their protestation to the contrary cannot hide that they are mainly dickering over price."

A key finding in Litt's report was his perception that the forty-two-inch Alaska Highway line with its design capacity of two point four billion cubic feet of gas per day was too small. "Alcan will have to expand its line by looping to carry any increased volumes efficiently and, if any additional volumes are projected for the early years, Alcan's line will be obsolete as the pipe is put into the ground. So will the cost estimates."

Three weeks after the Litt report, Trudeau was in Washington to address a joint session of Congress and to discuss with President Carter such items as energy and the proposed gas pipelines from the western Arctic. Returning to Ottawa, Trudeau told the House of Commons that Canada would try to reach its decision in time to meet the U.S. deadline of September 1st. He said that Canada "won't stand in the way of pipelines, as long as Indian land claims and environmental and economic problems are settled." NDP energy critic Tommy Douglas charged that Trudeau's assurances to Carter amounted to a "tacit understanding that Canada will proceed with a pipeline." He moved that Parliament, "express the opinion that the proposed construction of the Mackenzie Valley pipeline is not in the best interest of Canadian people," and urged that an Alaska Highway system be considered instead. The motion was defeated one hundred eighty seven to seven.

On the heels of the Litt report came the flood of petitions to the President urging approval of the Arctic Gas project. The Energy Task Force, representing the governors of eighteen midwest states, had already declared their strong support for the Arctic Gas route. The governors of the thirteen states making up the Appalachian Regional Commission wrote to urge that, "the President and the Congress recognize the merit of expediting approval of the Arctic Gas proposal for moving Alaskan natural gas to Appalachia and other areas of the country."

Thirteen United States Senators signed a letter to Carter claiming that the forty-five thousand page FPC record "provides conclusive evidence that the proposed Arctic Gas pipeline is the wisest of the three proposed delivery systems in terms of national energy needs, economic and environmental considerations." The National Association of Regulatory Utility Commissioners, comprising state regulatory authorities throughout the country, urged Carter to approve the Arctic Gas project because it "will afford the safest, most environmentally acceptable and most efficient method proposed to date for arctic gas movement." The United Auto Workers Union urged approval of Arctic Gas "in the national interest" claiming that the transportation of North Slope gas "must become one of the nation's highest priorities." (In Canada, the United Auto Workers had urged the government to cut off gas exports to the United States and delay a Mackenzie Valley pipeline in order to provide more time for a settlement of native claims.)

The environmental organizations, to be sure, were far from happy with the Arctic Gas route. Brock Evans, Washington director of the Sierra Club, told a House Public Lands sub-committee that environmental groups in both the United States and Canada were as one in their belief that the Arctic Gas route "is by far the worst choice," and that this line "should never be built." Brock said it would "invade and severely damage, if not completely destroy, one of the finest untouched wildernesses we still possess—and when there are viable alternatives to accomplish the same goal without so doing." He said there is no way to mitigate the environmental damage Arctic Gas would cause. "The Alcan route does appear to be the most environmentally sound," he said, even though it does raise some questions "that need to be resolved."

"Those who wish gas to come from Alaska to the U.S. have alternative ways to go without destroying the largest, most beautiful and wildest of all our wildlife refuges," Brock claimed. "But those who value the wilderness, the unbroken sweep of Arctic mountains to foothills to coastal plain to sea, have no other place to go."

Despite this environmental support, the prospects did not look too encouraging for the Alaska Highway line. This was made clear in a brief note from Stewart Udall, a Washington lawyer and former Secretary of the Interior, written to Ian MacDougall at York University one week after the Litt report:

Spoke to Blair today to express my dismay at FPC examiner's report—and the way the U.S. press almost completely ignored Alcan. I told him I thought they were getting clobbered in Washington—and expressed my personal feeling that they desperately need heavyweight representation if they intend to make a real fight for their proposal.

I further advised him that Schlesinger's [U.S. Energy Secretary James R. Schlesinger] two closest advisors [O'Leary and Freeman] are two of my closest 'energy friends,' and also conveyed to him my conviction that these three individuals will have a decisive influence on Carter's decision next summer.

Blair seems somewhat indecisive to me....

P.S. The FPC decision reflects the attitude that what the U.S. needs is the be-all and end-all, and what Canada needs and wants is a very secondary consideration. Right now this battle is being lost, but it is not too late to turn it around, in my opinion.

And turn it around the Foothills group was determined to do. If the United States wanted a pipeline larger than the forty-two-inch line they had proposed, then by God, they would file an application for a forty-eight-inch pipeline along the Alaska Highway route. This latest—the fifth—in the series of northern pipeline proposals led by Blair, would be an "express line" for the exclusive movement of Alaskan gas and would by-pass the use of existing facilities in British Columbia and Alberta which had been a feature of their forty-two-inch proposal so completely rejected by Judge Litt.

This newest Foothills proposal was first reported by Tom Kennedy, Calgary correspondent for the Toronto *Globe and Mail* and occasional contributor to CBC radio broadcasts. In his squeaky, high-pitched voice, Kennedy cackled the news on a CBC broadcast on February 16, 1977:

A pipeline proposal they hope to end all pipeline proposals is about to be announced by Foothills Pipe Lines and Alberta Gas Trunk Line. Don't ever say die when Bob Blair is around. The intrepid president of Alberta Gas Trunk Line who was dealt a severe blow only last week by the American authorities, is back today in the Arctic gas pipeline race with a vengeance. As a matter of fact, he's aiming a deadly torpedo right now at the establishment fat cats of Canadian Arctic Gas Limited, and after some 10 years of Bob Blair watching I would say that he may be down but not out... At this very moment his underlings are marching up Parliament Hill in Ottawa, and on

Capitol Hill in Washington, D.C., delivering the ammunition, which is a startling new proposal for a natural gas pipeline to move Alaskan gas.

This newest proposal, Kennedy reported, would provide the Americans with "their gas to their furnaces and homes and industries in the shortest possible time, by 1981." In addition, it "all at once would defuse" the native land claims issue in the Mackenzie Valley, by completely avoiding that region.

The Foothills group itself, however, seemed to be less enthusiastic than Mr. Kennedy about their own proposal. They had already stated that it would not be wise to build too large a pipeline to move Prudhoe Bay gas. In testimony filed the previous year with the Federal Power Commission, Ed Phillips had said that pipeline capacity greater than their planned forty-two-inch line would not be "prudent in the public interest... We believe that capacity of an average two point four billion cubic feet per day is more than adequate for the present, since it appears to us that deliveries from Alaska will not exceed that volume for some considerable time."

Now the Foothills group sounded almost apologetic in offering their proposal for a forty-eight-inch line with a design capacity one-third greater. In announcing this proposal at the Energy Board hearings the day after Tom Kennedy's report, it was characterized as "second best" by Foothills counsel Reg Gibbs. Despite this fact, Gibbs said, "it would still preserve for Canada and Alaska the obvious advantages of the Alaska Highway route over the Arctic Gas North Slope route." These advantages, he said, were "wholly Canadian ownership within Canada, and the use of existing experienced operating organizations in the persons of Westcoast and Alberta Gas Trunk." The *Anchorage Daily Times*, on the same date, quoted Blair as stating that the Foothills group had expanded the size of the Alaska Highway line because "that is what the U.S. seems to want. We still think we are right but we lost the argument ... Our preoccupation with prudence and conservatism went against the U.S. mood. In the current atmosphere of shortages and crisis, the U.S. dream is lots of gas from Alaska. If the U.S. wants an expensive system which clearly segregates gas and provides room for expansion, that's fine with us." Later Blair told the NEB that this line "will carry whatever Alaska gas is delivered to it up to its maximum capacity. The initial deliveries are advertised to be of the order of two billion cubic feet per day. I think they may turn out less."

Even this was still not the end of expanding the size of the planned Alaska Highway pipeline. By early 1978, regulatory authorities in Canada had decided that it ought to be a fifty-six-inch line with a design capacity in excess of four billion cubic feet per day. This was to provide room in the line for gas from the Mackenzie Delta in the event that the Alaska Highway line is ever connected with the Delta by the proposed seven hundred and sixty mile Dempster Highway line. There was no assurance that the Dempster line would ever be built, however. Blair and other Foothills spokesmen, as well as many others in the industry, have said that a pipeline up the Mackenzie Valley would be preferable to the longer and more costly Dempster route. And the National Energy Board itself found that a Mackenzie Valley route would be preferable to a Dempster route, if there are adequate gas reserves in the Delta area.

Never before has an application for a major gas transmission line received less regulatory examination than the forty-eight-inch Alaska Highway line. It occupied a bare two weeks of examination at the National Energy Board hearings by the time the hearings were completed at the end of April, 1977. In the United States, there were no public hearings on this latest proposal, the FPC hearing record having already been closed three months before. The only factor which allowed—and in fact required—the FPC to consider this proposal was the act which Congress had passed removing the FPC's decision-making power and putting it in the hands of the President and Congress. Because of this, the Federal Power Commission was obliged to consider the proposal, even without the benefit of public hearings, and report its findings along with the recommendations which were due by May 1st.

The FPC staff found the new Alaska Highway proposal little better than the old one. In their report to the Commissioners on April 8th, the staff concluded that the Arctic Gas route continues to be "the preferred delivery mechanism for North Slope natural gas." Once again, they described it as "vastly superior" to the El Paso and Alaska Highway alternatives, asserting that "the evidence is overwhelming." They concluded that the Arctic Gas project "which efficiently connects the Prudhoe Bay field to the Mackenzie Delta field, over the shortest route, becomes not only the least costly project, but the environmentally superior project."

Chapter 7
The Death of Arctic Gas

Depending on your viewpoint, the signposts marking the downhill journey of Arctic Gas are either painfully or joyously obvious. First came the issue of government financing, then the reports of the Federal Power Commission to President Carter, followed by the Berger, National Energy Board, and Lysyk inquiries.

Arctic Gas was unequivocal in asserting that financing its proposed line would require the assistance of the Canadian and United States governments, a political no-no. The Federal Power Commission was equivocal in its choice of an overland pipeline route; whichever route Canada wanted would be just fine. Justice Berger was adamant; no pipeline ever across the northern Yukon and not for ten years along the Mackenzie Valley, so as to permit the establishment of a native socialist utopia in the north. The National Energy Board was incapable of running counter to these political forces. The Lysyk Inquiry was accomodating and applied the *coup de grace*.

Any astute politician will confirm the folly of unequivocal stands on any issues short of motherhood, and even that is no longer safe. Arctic Gas paid the price for its political folly; they had clearly said that government financing was necessary. What Foothills was saying was, without close examination, not quite so clear.

One major point finally emerged, although it was not highly visible publicly: someone would have to help finance any pipeline built from the western Arctic, because the amount of money needed was simply greater than all the gas pipeline companies in both Canada and the United States would be able to borrow. Someone with far greater assets—"credit worthy parties" in the parlance of the financial community—would have to stand by to ensure the lenders that their loans would be safe, no matter what. Judge Litt, in his report, found that the pipeline companies "are not collectively strong enough to enter open-ended commitments to the lenders to complete the project or to guarantee repayment of the debt." He said that they "cannot guarantee, solely on the basis of their own balance sheets, that the project will be completed."

No one disputed this assessment—not the pipeline applicants, the financial community, nor any government agency. The question was, who should be the "credit worthy" parties? On whom should the bell be hung? Prospective candidates included the United States and Canadian governments, American gas consumers, as well as the State of Alaska and the oil companies, the owners of gas reserves who would benefit from the production and sale of Alaskan gas.

Backstopping of the debt by the taxpayers would require legislation in both Parliament and Congress. If U.S. consumers were on the hook, that, too, would require legislation in Congress. It would mean that if the project failed, the cost of any debt would be added to the bills of American consumers, even if they never received any gas from Alaska.

The position of Arctic Gas was that the United States and Canadian governments should stand behind the debt in proportion to the initial amounts of Canadian and U.S. gas that would flow through the proposed pipeline. Since, at least initially, the pipeline would carry mostly American gas, it was proposed that most of the guarantees would be provided by the U.S. Government. This stance was clear enough, and just as clearly controversial. Foothills' position on their two proposals—the Maple Leaf line for Delta gas and the Alaska Highway line for Prudhoe Bay gas—was more complex.

No business should ask the government to guarantee its debts, Blair argued. If Foothills was unable to secure the money to build its Maple Leaf pipeline, it would not ask the government for any financing help. Instead, it would simply ask that the government itself build, and own, a portion of the system, whatever portion Foothills might not be able to finance. As for the Alaska Highway line, if the United States Government wanted to help finance that system, that would be just fine.

In this respect, the Arctic Gas route entailed both an economic advantage and a political disadvantage. By providing additional volumes for transportation, Delta gas would help reduce transportation costs. It also involved the financial backstopping of the Canadian government. On the other hand, a line moving only Alaskan gas, such as the Alaska Highway system, would remove this onus from Ottawa and leave it with Washington, a much more appealing solution in Canada, except that such a line would not provide any gas for Canada.

"We would discourage the asking, and would criticize the asking of a governmental guarantee for a project to be owned and to be profitable in the private sector," Blair told the National Energy Board. "If there were a situation in which a pipeline was considered in the public convenience and necessity, but could not be financed in the private sector... our recommendation would be that a Crown corporation be employed for the portion of the pipeline that could not be financed in the private sector."

"With the government guarantee situation, what you get is a free ride for the private sector, which can put up some money and get a full utility return to capital, or return to common equity for its money while the government takes the risks," Blair said. "I guess any of us probably... would love to have a certain fifteen or sixteen per cent return to common equity after income tax with the government picking up all the down side risk if anything goes wrong... We just do not think anybody is going to be in business, asking for that kind of investment, and we think it is wrong to seek such a guarantee."

Blair did not feel the same way about the U.S. government. "We already know that the United States portion of the project [the Alaska Highway line] has decided that the whole project can only be accomplished if there is a large United States Federal Government backstop," Blair said. "Since all that is in place for the U.S. end, we may as well have the advantage of it over the Canadian end, too, because the whole project is for the service of the United States' purpose, and we do not mind accepting any financial strength that we can get from the United States Federal Government to carry out such a project. It is in our own self-interest to accept everything that comes our way... We think politically and conceptually it is quite all right for the United States government to back up a project for U.S. purposes."

"You can do anything with the U.S. Government guarantee," Trunk Line vice-president Robert Pierce later told the Energy Board. "I think you can make a non-viable project into a viable project just by virtue of the U.S. Government and its tremendous tax base."

While the Foothills group looked initially to the American government as the credit worthy party, this was later modified after the U.S. Treasury Department expressed some strenuous objections. Foothills then looked to U.S. consumers for financial support. This was to be accomplished by a tariff approved by legislation, under which the American gas utilities, and thus their customers, would start paying for the proposed pipeline by a certain date, whether or not the pipeline was completed by that date, or whether it was ever completed.

W. A. Davidson, first vice-president of Loeb Rhoades & Co. Inc., a major New York investment house and one of Foothills' financial advisor firms, explained the approach to the Energy Board:

> We have always maintained that you could get the financing done with the [perfect tracking] tariff if the tariff is set in place and locked in place with legislation. We have also said you can get the financing done with a federal guarantee if the federal guarantee is authorized by legislation. It has become Northwest's [Northwest Pipeline Company] position that they would rather go with legislation for the tracking as opposed to legislation for the guarantee. As you know, the Treasury Department in the United States has adamantly opposed any suggestion of a federal guarantee . . . I do not personally have any view that one of these pieces of legislation would be much more controversial than the other. I think either one would be.

Arctic Gas agreed that either approach would work, but did not think it very likely that Congress would pass legislation putting the consumers on the hook. "Our position has been that it is quite possible that if we got all of the measures which are involved in perfect tracking [of tariff charges], the project could be financeable," Bill Brackett, vice-chairman of Alaskan Arctic Gas, told the Energy Board. "My analysis of the political situation in the United States is that there is no feasible possibility of legislation to require state regulatory agencies to pass on those charges." Ross LeMesurier, vice-president of Wood Gundy Limited, one of the Arctic Gas financial advisor firms, explained it to the board this way: "If we could get some fool-proof mechanism by which every customer of natural gas, ranging from General Motors to Mrs. Smith, to cheerfully—well, they don't have to do it cheerfully—but to pay their bills every month, and we had a fool-proof mechanism whereby this would work ... then we would probably not need the government support. But ... it seems impractical."

Thus, the major difference in the financing approaches of Arctic Gas and the Foothills Alaska Highway line was that one opted for legislation to put the U.S. taxpayer on the hook and the other opted for legislation to put the gas consumers on the hook, with Arctic Gas also involving the Canadian government because it would move Canadian gas.

In the end, approval of the Alaska Highway line completely rejected both of these approaches. Approval was premised on the assurance of the Foothills sponsors that they could finance the project in the private sector, given financial support from the State of Alaska and the oil companies. If it seemed logical to expect the State of Alaska to financially support the move of Alaskan gas because it has a royalty interest in the gas, then by the same reasoning it would seem just as logical for the Government of Canada to support the movement of Mackenzie Delta gas. The State of Alaska, however, as well as the oil companies, were just as adamant that they would not support the financing. The State of Alaska did later agree to underwrite $1 billion in non-taxable bonds, an amount that fell far short of meeting the needs of the project.

If the Arctic Gas plan to seek financial backstopping from the two federal governments was not exactly greeted with enthusiasm, neither did it appear to be fatal. No one else appeared to have come up with any better idea of how to borrow the money.

The first clear indication that Arctic Gas was headed for trouble was the report on May 2, 1977, of the Federal Power Commission to the President. The four commissioners, in a six-page letter accompanying their report, expressed clear preference for a pipeline across Canada to deliver Alaskan gas, but they split two-two on which pipeline it ought to be. "Based on today's circumstances, reasonable men can disagree on the right course of action," they wrote to the President. FPC chairman Richard L. Dunham and vice-chairman James G. Watt (both Republicans) favoured the Alaska Highway line; commissioners Don S. Smith and John Holloman, III (both Democrats) favoured Arctic Gas. It was a dramatic change from the findings of Judge Litt and the FPC staff who had declared Arctic Gas to be "vastly superior."

A key factor was the FPC's assessment that there would be less gas available from the Delta than Judge Litt had indicated, and less than Arctic Gas had projected. This implied that the design capacity of the Arctic Gas system might not be fully utilized, reducing the transportation cost advantage which it might otherwise offer. They concluded that the estimated six to seven trillion cubic feet of gas reserves found in the Delta were not likely to "increase significantly in the near future. While Arctic Gas has projected that Mackenzie Delta reserves might grow to support up to two and a quarter billion cubic feet per day deliverability, we believe it unlikely to exceed one billion cubic feet per day in the near term." Even with this limitation, they still found that the Arctic Gas route offered the lowest transportation costs. Costs for the Alaskan Highway line were estimated at four per cent higher, and for El Paso at forty per cent higher.

While the FPC gave the economic edge to Arctic Gas, it gave the environmental edge to the Alaska Highway route. "Each system will have some adverse environmental impacts," the commissioners wrote. "We believe all of these impacts to be acceptable, given precautionary measures. Arctic Gas would involve crossing the Arctic National Wildlife Range and other lands now little used by man. The other projects would generally follow existing utility corridors— a distinct environmental advantage."

In Canada, the split decision by the FPC was widely regarded as a signal that the choice was really up to Canada. Since it was Canadian land that was largely involved, it seemed like a reasonable position. It also had the effect of relieving the Carter Administration of a tough political choice. No matter which route Carter chose there was bound to be some political opposition in the United States, if not from the midwest region, then from Alaska. By leaving it up to Canada, the Carter Administration escaped that problem.

Later, I spoke to Richard Dunham in Washington. He had by then retired from the FPC and joined a management consulting firm where he said he was busy "intellectualizing." He confirmed that the split decision clearly did signal that the choice was up to Canada. All four commissioners were prepared to support whichever route was selected by the administration, Dunham said, and the administration could wait for the word from Canada before making up its mind.

Dunham indicated that since Congress had left the final decision up to the President, there was concern that the FPC should not reduce the options available to the President. He also noted that there were factors involved in the final decision, such as defense matters, which were beyond the purview of the FPC.

Another factor, said Dunham, was that the economics between the Arctic Gas and Alaska Highway routes did not seem great. "When the objective factors [economics] are close, then the subjective factors [environmental and social], which are matters of individual judgment, assume increasingly greater importance." In Dunham's view, the subjective factors favoured the Alaska Highway route.

Had the Delta reserves been greater, the economic advantages of the Arctic Gas proposal would have been much greater, and its chances for approval much better, Dunham explained. He said he figured that eleven to twelve trillion cubic feet of Delta reserves represented the "window" which would have favoured Arctic Gas, while about twenty trillion cubic feet would have permitted an all-Canadian line from the Delta. (Former National Energy Board chairman Marshall Crowe had mentioned exactly the same point to me, using the same numbers, when I interviewed him in Ottawa.)

Dunham talked about the small size of the gas fields found in the Delta and the disappointing exploration trends. He seemed surprised when I mentioned that the Taglu field was the second largest gas field in Canada; that Delta reserves amounted to fifty billion cubic feet of gas for each exploratory well drilled, compared with less than three billion cubic feet for each exploratory well drilled in western Canada; that the Geological Survey of Canada had estimated that the Delta and Beaufort Sea area had a larger potential for the discovery of new gas reserves than any other region of Canada.

Regardless of this, Dunham claimed that the split decision of the FPC provided the best prospects that Arctic Gas could have expected. Had all four commissioners endorsed Arctic Gas, he said, it would have brought screams of protest in Canada that the United States was dictating what Canada should do, without regard to Canadian interests.

THE BERGER REPORT

If there was any uncertainty about the effects of the FPC recommendation to the President, there was none about the recommendation of Justice Berger to the Minister of Indian Affairs and Northern Development. The Berger report came one week after the FPC report. Of the three principal inquiries, Berger's was the only one not charged with a responsibility for finding whether or not a pipeline should be built, or when. Yet it was the most emphatic in its views: never across the northern Yukon and not for ten years along the Mackenzie Valley.

An interesting aspects of the Berger report is that nowhere does it contain any finding that a gas pipeline across the northern Yukon to link the Prudhoe Bay and Delta gas reserves would, if built in the manner proposed, result in intolerable environmental effects. Predictions that the wilderness values of this region would be ruined, and the wildlife resources decimated, were predicated on the assumptions that Arctic Gas would not be able to construct its pipeline during just the winter months, and that this line would be followed by an oil pipeline, a year-round gravel road, and accelrated petroleum exploration activity.

"I am not persuaded that Arctic Gas can meet its construction schedule across the northern Yukon," Justice Berger wrote. "Should this occur, there is a likelihood of cost overruns, of construction being extended into the summer, or even a permanent road being built to permit summer construction. The environmental impact of a change to summer construction would be severe."

The same environmental concerns were not found to be present in the Mackenzie Valley. "I have concluded that it is feasible," Justice Berger wrote, "from an environmental point of view, to build a pipeline and establish an energy corridor along the Mackenzie Valley." Native people, however, fear that a pipeline will bring "an influx of construction workers, more alcoholism, tearing of the social fabric, injury to the land, and the loss of their identity as a people," Berger wrote. To avert such tragedies, native people seek a settlement of their claims which will address their concerns about "land and land use, renewable and non-renewable resources, schools, health and social services, public order and, overarching all of these, the future shape and composition of political institutions in the north."

"In my opinion, a period of ten years will be required in the Mackenzie Valley and western Arctic to settle native claims, and to establish new institutions and new programs that a settlement will entail. No pipeline should be built until these things have been achieved." If a pipeline were built immediately, Berger reported, "there is a real possibility of civil disobedience and civil disorder that—if they did occur—might well render orderly political evolution of the north impossible, and could poison relations between the Government of Canada and the native people for many years to come."

Berger looked askance at the Arctic Gas proposal as involving a risky incursion of American interests. He described the Arctic Gas pipeline as "a land bridge for the delivery of Alaskan gas across Canada to the lower 48 . . . an energy lifeline for the United States." He claimed that the United States "cannot be expected to be as concerned as Canada with the seriousness of the social and environmental impact of a pipeline along its route . . . The risk is in Canada. The urgency is in the United States." The pipeline, he said, would "cross lands that are claimed by Canada's native people, a region where the struggle for a new social and economic order and political responsibility is taking place." By blocking a line to link Prudhoe Bay and Delta gas, Canada would be free to build its own pipeline from the Delta. Thus, "we can build a Mackenzie Valley pipeline at a time of our own choosing along a route of our own choice." How much Berger's view of a pipeline route across the northern Yukon might have been influenced by these political, rather than environmental concerns is a matter of conjecture. Curiously, he expressed no similar doubts about the intrusion of United States interests and the dangers of a land bridge related to a pipeline to carry American gas along the Alaska Highway and across the southern Yukon. At the same time, he did not hesitate to point out the claimed environmental benefits of the southern route.

Perhaps never before has the release of a government inquiry report been attended with so much news media coverage as the release of the first volume of Berger's report. It was tabled in the House of Commons at 3 pm, May 9th, 1977. Scores of news reporters had already been poring through the report since early that morning in a lock-up session similar to the manner in which reporters gain early access to the federal budget. CBC had been working for weeks on a prime-time, hour-long television special to be broadcast on the day the report was released. Included in this special was a live-broadcast, relayed by satellite, of a public meeting in Inuvik where northern residents voiced their initial reactions to Berger's recommendations. The Inuvik hall was packed to capacity. Justice Berger sat in the CBC's Ottawa studio that evening, watching the relay of this northern meeting. He appeared visibly startled as most Inuvik residents, including most of the many native people present, expressed their strong opposition to a ten-year delay in construction of a Mackenzie Valley pipeline.

In Vern Horte's mind there never was any question that Berger's report would be bad news for Arctic Gas. The only question was how bad. It managed to fulfil all his apprehensions.

Horte had agreed to be available to news reporters following the release of the report. A meeting room had been arranged in the Four Seasons Hotel. It was nearly 4 pm before Horte saw a copy of the report. By 7 pm he was ready to face the reporters and their barrage of questions. As the reporters and cameramen jammed in, it soon became apparent that a room twice as large should have been provided.

"We disagree fundamentally with the Berger report's conclusions," Horte stated. "I think if Canada were to accept his conclusions, it would mean obviously the cancellation of our project. Beyond that it would also mean that Canadians would not be able to enjoy the possible advantages of a joint pipeline and the availability of its natural gas in the Mackenzie Delta."

Horte estimated that if it were later possible to build a pipeline to move just Delta gas, the cost of transportation during a twenty year period would be as much as $6 billion greater than if it were moved in the same pipeline with Alaskan gas. "That in my opinion would be the consequence to Canada if the Berger recommendation of a ten year moratorium were accepted."

He also claimed that for northern residents, economic activity and the provision of more jobs "is absolutely essential. I am not saying that land claims are not also essential. But the one without the other, in our opinion, will really not solve the problem."

"Mr. Justice Berger was not charged with looking at the total Canadian public interest, or making a decision as to whether or not in the final analysis Canada should have a pipeline," Horte said. The National Energy Board would have to consider broader aspects of the Canadian interest in its report to the cabinet. "The Government of Canada will have to make its ultimate decision on the basis of that total input. I think it would be wrong to consider the Berger report to be the final answer."

But the Berger report was, in effect, the final answer. The reports

107

of the National Energy Board and the Lysyk Commission merely confirmed it.

THE NEB REPORT

The offices of the National Energy Board are located in Ottawa at 473 Albert Street in the Trebla Building. That is Albert spelled backwards. At least in the view of Arctic Gas, it was not the only thing that the National Energy Board got backwards. There was also the board's decision on the applications for a natural gas pipeline from the western Arctic.

The board had completed its public hearings on the applications on May 12th, after two hundred and eleven hearing days that filled thirty-seven thousand, four hundred and fifty-five pages of transcript, not counting the earlier aborted hearings chaired by Marshall Crowe. Eight weeks later, at 5 pm, Monday, July 4th, the applicants and intervenors assembled for the final time in the board's hearing room on the ninth floor of the building spelled backwards.

All the big guns were there to listen to the board's hearing panel "announce its decision and give reasons." There was John McMillian from Northwest Pipeline Company, Blair from Trunk Line, Wilder and Horte from Arctic Gas, representatives of the special interest organizations, reporters, and lawyers by the case load. The bench where the hearing panel sits, like a judge's bench in a courtroom, is at the far wall opposite the entrance doors. Seating throughout the rest of the room is U shaped. In the centre of the U is the long table at which witnesses normally present their evidence, but on this day this area was filled with extra seating to accommodate the capacity crowd. At a given signal, everyone stood, just as in court, while the hearing panel entered in procession, first Geoff Edge, then Jack Stabback and finally Ralph Brooks.

One thing that each of the hearings considering these applications had in common is that each was unprecedented. They said so themselves. "The hearing was unprecedented," Jack Stabback said from the bench, "not only in its length, but in terms of its magnitude and pervasiveness of its issues and of its importance to Canadians and Americans alike. It is perhaps fitting, although coincidental, that this decision [with a capital D in the transcript] should be announced on a date almost immediately following Canada Day, and on American Independence Day."

The audience sat for an hour as the panel members took turns reading various portions, first giving the reasons, then finally the decision. They sat with poker faces, Reg Gibbs staring unblinkingly, his face as immobile as granite. When Stabback had finished reading the final portion, John McMillian was grinning like a school boy; Bob Blair had lost his constant stern expression; Wilder, Horte and the other Arctic Gas representatives were stunned and grim.

By the time the hour was over, the board had rejected both the application of Foothills for its Maple Leaf line (which, in any case, Foothills had withdrawn), and the application of Arctic Gas to move both Prudhoe Bay and Delta gas. In pace of these, the board had developed its own proposal for pipelines to move gas from both sources across routes that had not been applied for and with its own version of ownership arrangements, then offered this package to the Foothills group.

The board's proposal called for an Alaska Highway line to move Alaska gas, but with a large diversion of the route to pass by way of Dawson City. This would increase the length of the line carrying Alaskan gas, but would reduce the length of a possible future lateral along the Dempster Highway to connect Delta gas fields with the Alaska Highway line. The effect of this, the board estimated, would be to increase the cost of moving Alaskan gas by six cents per thousand cubic feet, but reduce the cost for Delta gas by twelve cents. "The six cents per mcf [thousand cubic feet] is small compared with the additional 30 cents per mcf estimated by the FPC if the El Paso project were selected," the board stated.

The fact that there had been no significant engineering, economic nor environmental studies of these routes did not appear to bother the board, because neither had there been any substantial studies of the Alaska Highway route. "Recognizing that the amount of engineering design work and environmental and socio-economic studies and planning needed to meet final design requirements on the Alaska highway route would be substantial in any case, it is the board's opinion that the Dawson diversion would not significantly alter the existing proposal nor the construction schedule." The board also found that Foothills "has not had sufficient time to make the necessary terrain analysis and to finalize its design in respect to frost heave and thaw settlement" in the Yukon portion of the Alaska Highway line, and that the cost of this section might be as much as $1.8 billion dollars or twenty per cent more than estimated.

The board's record in developing pipeline proposals fared little better than that of the applicants. In their subsequent negotiations, both the Canadian and American governments rejected the board's recommended route, returning to the route that the Foothills group had applied for, but starting all over again on studies as to what diameter of pipeline ought to be installed. It almost seemed as though everything had moved back to 1969 to start again from scratch.

The board said it was "seriously concerned" about the ability of Trunk Line and Westcoast to finance the Canadian sections of the Alaska Highway line, particularly since "there is considerable evidence that costs could be significantly overrun." This was compounded by a concern that Alberta Gas Trunk Line (Canada) Limited "might be unduly fettered" as a wholly-owned subsidiary of AGTL, "a company exposed to risks in the chemical industry and regulated by provincial authorities with regard to its pipeline activities."

The solution advanced by the board was that Foothills Pipe Lines should own one hundred per cent of the pipeline section in the Yukon, and fifty-one per cent of the sections in British Columbia, Alberta and Saskatchewan, with the remaining forty-nine per cent of the various segments owned by Trunk Line, Westcoast, Trans-Canada and Alberta Natural Gas.

The board suggested—and the government later agreed—that Foothills should be required to begin an immediate feasibility study of the contemplated Dempster pipeline link and should file applications for such a system by July 1, 1979. But whether such a line should ever be built was left uncertain. If new reserves discovered in the Delta exceed fifteen trillion cubic feet, it will probably be more economical to transport them by a pipeline along the Mackenzie Valley than by the longer route across the Dempster, in the board's opinion.

The board also found that Delta gas reserves "appear to be one of the most economic new sources of energy" available to Canada, "and probably the most attractive one." It found that the Arctic Gas route could connect these reserves at the least cost and with the greatest economic benefit for Canada. But in rejecting the Arctic Gas bid, the Energy Board gave way to the concerns raised earlier by Justice Berger. It found the Arctic Gas route "environmentally unacceptable," because it involves "impacts which could not be

110

avoided, which could not be accepted, and for which mitigative measures are unknown or uncertain of development." It concluded that more time is needed to settle native land claims before building a pipeline in the Mackenzie Valley, while in the Yukon "the native population which would be affected is smaller." And it categorically rejected the Arctic Gas application for federal government financial backstopping.

In its assessment of relative economics, the board concluded that the Alaska Highway and Dempster routes, even though they involve an extra thousand miles of pipeline, would result in only "slightly" greater transportation costs than the Arctic Gas route. The reason for this is that the supply volumes used by the board showed the Alaska Highway line operating at ninety-seven per cent of its then smaller design capacity, while the higher pressure Arctic Gas line would be operating at only seventy per cent of its design capacity. The board did not indicate what the cost comparison would be with the proposed Arctic Gas line operating at full capacity, because it assumed that there would not be sufficient Delta gas to achieve this. Nor, alternatively, did the board indicate what the economic comparison would be for a smaller capacity pipeline along the shorter Arctic Gas route operating at design capacity.

In reading the board's decision, Stabback said: "There can be no doubt that the reserves found on land in the Mackenzie Delta area have been disappointing—only five point three trillion cubic feet of established reserves as estimated by the board—and the board cannot rely on the expectation that there will be large finds of gas in the near future in the Beaufort Sea."

In the three-volume report accompanying its decision, however, the board does rely on the discovery of twenty-nine trillion cubic feet of gas in western Canada during the next twenty years in calculating the gas supplies available to meet Canadian needs. The board does count on new gas discoveries in calculating the amount of surplus gas available for export, but apparently does not count on new discoveries in evaluating alternative transportation routes.

Within four months of the board's decision, Dome Petroleum reported that it had made four gas discoveries at the first four exploratory wells drilled in the deep offshore portion of the Delta-Beaufort Sea basin. The discoveries provided dramatic confirmation of the earlier Geological Survey of Canada assessment that this region offers the largest potential for new gas reserves in Canada.

The final inquiry to report was the Lysyk Commission. It was asked to provide Indian and Northern Affairs Minister Warren Allmand with a preliminary assessment of the social and economic impacts of the proposed Alaska Highway pipeline and the views of Yukon residents. The three-person inquiry was headed by Kenneth Lysyk, Dean of Law at the University of British Columbia, and included two Yukon residents, Edith Bohmer representing the Council for Yukon Indians, and William Phelps, representing the Yukon Territorial Council. Appointed on April 19, 1977, the Lysyk Inquiry issued its report in less than four months, on August 1st, after holding twenty-two days of formal hearings in Whitehorse and twenty-seven days of informal hearings in Whitehorse and sixteen other communities. By contrast, Berger's preliminary report required more than three years, and his final report nearly four years.

Timing was not the only difference in the reports of these two northern inquiries. Where Berger found conflict, the Lysyk Inquiry found compromise. Berger found the north to be "a region of conflicting goals," with the conflict focused on the pipeline which "represents the advance of the industrial system to the Arctic." Where Berger was determined that native peoples should not be losers, the Lysyk commission seemed determined that everyone should be winners.

"It is clear to us that to proceed immediately with the construction of this pipeline ... would run counter to the aspirations of many Yukoners and that it would disappoint them bitterly," the three members of the Lysyk Commission wrote to Allmand. "It is equally clear to us that many Yukoners regard properly controlled economic development not only as inevitable but also as desirable, and they see in the proposed pipeline an opportunity to improve economic and social conditions in the Yukon. A lengthy or indefinite moratorium on the construction of the pipeline would seem to be, for practical purposes, a decision against building it through the Yukon."

They concluded that "the proper solution to this problem seems to us to lie in flexibility and compromise. Middle ground must be found that will be acceptable to most, if not all Yukoners."

Had Berger brought as accommodating an attitude to his inquiry as the Lysyk Commission, many Northwest Territories residents are now convinced that there would have been no ten year moratorium on a Mackenzie Valley pipeline.

112

As Arctic Gas surveyed its abandoned hopes and plans, there was at least one consoling thought. Things could have been worse. It might have been Arctic Gas that got stuck with the Alaska Highway pipeline.

PART TWO
PUBLIC ISSUES

"It is always with the best intentions that the worst work is done."

Oscar Wilde

Chapter 8
Marshall Crowe and the National Energy Board

Ever since it was launched in 1959, Canada's National Energy Board seems to have been swimming in hot water. That water could get hotter if the pipeline selected by the board to transport Mackenzie Delta and Alaskan North Slope gas is substantially deferred or frustrated.

It was the Energy Board, not the Canadian government nor the United States government, that selected the Alaska Highway route. In the United States, the Federal Power Commission very deliberately left the U.S. decision to President Carter and his administration. In Canada, the National Energy Board, in effect, left the federal cabinet with no say in the matter. Energy Minister Alastair Gillespie, for one, has been critical of the board for that.

So if the Alaska Highway line flounders, it will be the Energy Board which will face many of the questions. Questions about how the board arrived at its decision. Questions about the former board chairman, Marshall Crowe, whose decision to chair the NEB hearings upset the entire hearing schedule, and who later made no secret of his preference for the Alaska Highway route.

Controversy is nothing new for the National Energy Board. The New Democratic Party has long accused it of being the handmaiden of the oil industry, all too eager to acquiesce to U.S. interests. Former NDP energy critic Max Saltsman once referred to it, in the House of Commons, as "Canada's National Monkey Board." When the board approved some six trillion cubic feet of gas export sales in 1970, it was accused of a sellout of Canadian resources. The oil industry seemed hardly any happier, upset that the board had rejected further export applications of some four trillion cubic feet. Environmental and conservation organizations have felt that the board has not given sufficient consideration to their views. The inability of the public to effectively participate in the board's public hearings without incurring heavy costs to cope with the exhaustive and arcane arguments of experts has been another sore point for some.

Even the very functions the board is obliged to carry out have caused concern. Under the act creating the board, it is required to act both as an advisor to the government on matters of energy policy and as a regulatory tribunal with considerable decision making powers. The concern has arisen over whether or not it is possible for the board to avoid being influenced in its regulatory decisions by the policy advice which it offers the government.

To be sure, the board has been faced with some almost impossible circumstances. Nothing short of a total ban on all new energy development could have avoided incurring the wrath of at least some environmental organizations during the past decade. Public resentment of rapidly rising energy prices, charges of a "phony" energy crisis, mistrust of the big oil companies, and conflicting and confusing energy forecasts have all inevitably rubbed off onto the board. There is nothing the board could have possibly done to avoid all of this. If it has not always been on the mark in predicting the imponderable energy trends of the past decade, who has? The fact is that in this regard, no one has done any better than the NEB, nor even as well.

The Energy Board was born out of controversy. When John Diefenbaker led the Tories into office in 1957, after twenty-two years of uninterrupted Liberal rule, it was in the wake of controversy over the previous government's handling of the TransCanada PipeLines issue in Parliament, the financing of Westcoast Transmission, and the price of gas export sales to the United States. The report that year of the Royal Commission on Canada's economic prospects, headed by Walter Gordon, recommended establishment of an energy authority to advise the government "on all matters connected with the long-term requirements for energy." The Conservative government appointed another Royal Commission, headed by Henry Borden, to deal specifically with energy matters, such as the regulation of pipelines and the export of gas and oil.

The new government was anxious to avoid becoming trapped in the type of pipeline controversy that caught the Liberals and saw the establishment of an independent authority as a means of accomplishing this. The Borden Commission was specifically asked to make recommendations on "the extent of authority that might best be conferred on a national energy board to administer, subject to control and authority of Parliament, such aspects of energy policy coming within the jurisdiction of Parliament, as it may be desirable to entrust to such a board." After the Borden report was submitted, Parliament passed the National Energy Board Act in 1959.

Charges that the Energy Board is a captive of the interests of the oil and gas industry are refuted in a 1977 study prepared for the Law Reform Commission of Canada by Alastair R. Lucas of the University of British Columbia and Trevor Bell. Instead, the authors suggest that the board may be captive to political influence, arising from its dual function as both a regulatory and advisory body.

This report was written before the board's decision on northern pipeline applications. The authors suggest, however, that in making decisions of this type, the board is acutely aware of and influenced by political implications for the government.

The National Energy Board, the authors declare, "is an important tribunal whose decisions directly or indirectly touch the lives of every Canadian." When it was first established, the board was virtually the only agency providing the government with advice on energy matters. The later establishment of the Department of Energy, Mines and Resources brought some overlapping of the advisory functions. "Inevitably, some competition for the Minister's ear seems to have resulted," the authors state. Despite this, the board, either acting in its own capacity or by serving on interdepartmental study committees, continues to play a key role in shaping national energy policy, according to the authors. At times, decisions by the board are, in effect, national policy decisions. The "board and cabinet appear to work as a policy-making team," the authors state. "While not co-equal, the board may be likened to a very capable, experienced and highly informed senior aide whose advice can be regarded as highly reliable."

The concerns that this raises are set out clearly in the Lucas and Bell report:

> Good advice is politically viable advice. Successful advisors have understandably developed acute political sensitivity and the ability to screen out elements of their advice that may be affected by partisan or 'positional' considerations or be out of tune with a Minister's personal preferences given past statements and performances.

However, success in mastering the advisory role may, in a National Energy Board member, profoundly affect the member's ability to carry out adjudicative regulatory functions in a completely impartial manner. His instinct could well be for the 'politically viable' decision, if more general policy considerations were involved. In any event, he would probably already have played a role in providing advice to government on these more general policy issues. Sometimes... the subject of the application is itself a previously decided element in general government policy.

It is not surprising that concern has been expressed by representatives of many of the interests participating in board proceedings about the very nature of adjudication that allows it to be influenced by the advisory activities of board members. It has been suggested that the NEB is likely to make decisions based on members' 'political insights' even when there is no suggestion of direct or indirect consultation with the Minister on a particular application. Furthermore, some observers believe that board members probably see nothing wrong with this situation because of a tendency to judge their behaviour in terms of public service policy advisors rather than 'quasi-judicial' decision-makers. All agree that wide-spread suspicions generated by the combination of functions, whether well-founded or not, are extremely damaging to the NEB's credibility as an adjudicator. This in turn can reduce public as well as industry confidence in the Board and impair its ability to exercise its statutory mandate effectively.

NEB critics other than Lucas and Bell have raised almost opposite questions. There is little doubt that not only Energy Board decisions, but also those of other regulatory agencies at times determine matters of national policy, often involving factors of political concern. Should regulatory agencies, acting independently of the government, be free to make decisions with broad national policy and political implications? Or should this responsibility rest with the elected government? One side of the argument is that policy making is the responsibility of those who were elected and are accountable to Parliament, rather than appointed and unaccountable officials. The other side of the question is that the hearings of regulatory bodies provide a means of public participation that is not otherwise available and also provide more stability than elected politicians who change their political priorities regularly and rapidly.

Stephen Duncan reported in the *Financial Post* on April 2, 1977, that early in the year the federal government was already working on draft legislation that would give cabinet more direct, and more open, control over the Energy Board and other agencies. "Cabinet wants greater control over the regulatory agencies, and it wants the agencies to have less power to make policy," Duncan wrote. He quoted a spokesman in the Privy Council office as expressing concern that, "these agencies have gradually increased their power to the point where they have enormous influence on policy. What the government is trying to do is bring this policy power back to ministers who are accountable to Parliament."

Whatever the final outcome of these debates, there is no doubt that the board's decision on northern pipeline applications was almost universally seen as politically influenced. Typical of this perception was the commentary on CTV television by Tom Gould on the day the board's decision was announced. Gould said that the board's report "would appear to be the answer to our federal government's prayers."

"Ottawa is under tremendous pressure on this matter of a natural gas pipeline," Gould commented. The United States wants to move its Alaskan gas by pipeline across Canada and "the Canadian government would like to accommodate the Americans." However, "the political problems up to now have appeared insurmountable. The Berger Commission hearings have advanced warning that a pipeline running across the Mackenzie Delta and down the Mackenzie Valley would become a symbolic battleground, a great showdown between those who support development and those who demand conservation... Although the National Energy Board is not a political body, it is closely attuned to the political realities of our federal capital. It has given our government an out," with the Alaska Highway route.

Politically-influenced or not, the board rejected the Arctic Gas route for the stated reasons of protecting environmental values and the opposition of the Indian Brotherhood of the Northwest Territories. What the board completely failed to provide was any meaningful assessment of the costs involved in avoiding the claimed environmental and social impacts. Costs not only in terms of dollars, but in terms of employment, national economic interests, and energy conservation. An enormous volume of energy will be consumed in moving the gas through the additional thousand miles of pipeline selected by the board. The only unqualified evaluation offered by the board was that a pipeline along the Alaskan North Slope and the Mackenzie Valley, operating at seventy per cent of design capacity, would offer only slightly cheaper transportation than two pipelines a thousand miles longer along the Alaska and Dempster highways, operating at ninety-seven per cent of design capacity. A comparison on this basis certainly provides no meaningful evaluation to determine the optimum costs and energy use. How is it possible to determine if the claimed environmental and social advantages of the Alaska Highway and Dempster routes are worth the cost, if the cost is not known? In this regard, the National Energy Board offered the Federal Government no useful advice at all.

The mantle of controversy hanging over the Energy Board fell soon enough on the shoulders of its chairman, Marshall Crowe, one of the most trusted and respected policy advisors in the ranks of the federal civil service hierarchy. The debate about whether he should chair the pipeline hearings was the first but not the last controversy Marshall Crowe encountered at the National Energy Board. Hardly had the dust settled from this than the Toronto *Globe and Mail* (September, 1976) splashed a story across page one reporting that Crowe had been the guest of Panarctic Oils Ltd. on a four-day fishing trip in the Arctic Islands. (Later Crowe told me that it "was not a fishing trip, it was a tour of Panarctic drilling sites, during which we did some fishing.") Others on the tour included Charles Hetherington, president of Panarctic and John Holding, president of Polar Gas (partly owned by Panarctic), a company preparing its own application to the board for a gas pipeline from the Arctic Islands. NDP leader Ed Broadbent and Conservative Energy critic James Gillies both accused Crowe of a serious error of judgment. Broadbent said that government officials "should ensure that their actions are beyond question . . . they should not accept the gift of a fishing trip by such a corporation." Gillies said, "I find it astonishing he was there at all."

More astonishing was Crowe's espousal, after he had been disassociated from the hearings but was still chairman of the board, of the merits of the Alaska Highway route. On April 26, 1977 while the board hearings were still in progress, Crowe spoke to some thirty representatives of Canadian financial and banking institutions at a luncheon held in the Toronto boardroom of MacLeod, Young, Weir & Company Limited, one of the firms acting as financial advisors to Foothills. Crowe presented to this financial audience his views on a wide range of energy matters, including those on competing applications for a northern gas pipeline. He explained that since he had been divorced from this matter by virtue of the Supreme Court ruling, he was free to express his own personal observations.

Crowe described the report of Judge Litt's FPC hearings as, "just another input" into the final decision. He said he doubted that the final FPC report to the President would come out very strongly in favour of either the Alaska Highway or Arctic Gas proposals. (Six days later, the four FPC commissioners, in their report of May 2 to U.S. President Jimmy Carter, were equally split between the Alaska Highway and Arctic Gas routes.) Crowe suggested that the Berger report, to be released May 9, would strongly militate against a Mackenzie Valley pipeline. He said that Berger's report would not be a factor in the board's decision, but it would carry great political weight. He suggested that approval of the Alaska Highway line would involve fewer political difficulties than the Arctic Gas route. The Alaska Highway line, he suggested, would avoid political problems which would result from the Berger report.

Crowe said that the Arctic Gas project would make sense if there were about ten trillion cubic feet of gas reserves in the Delta, but that it was a questionable proposition given the current level of four trillion cubic feet of reserves. (This was substantially less than any other estimate of Delta reserves at that time or subsequently.) Crowe observed further that more gas had been found in Alberta in 1976 than had been found to date in the Delta (ignoring that more than two thousand exploratory wells were drilled in Alberta in 1976 compared with a cumulative total of less than one hundred and fifty in the Delta area). Finally, Crowe suggested to his audience that the gas reserves in the Delta could be marketed by means of building a pipeline along the route of the Dempster Highway to connect with the Alaska Highway line; this is exactly what the board recommended ten weeks later.

The following month, Crowe spoke in New York to a similar group of financial representatives at a luncheon sponsored by Loeb Rhoades & Co. Inc., one of the American financial advisors for the Foothills Alaska Highway project.

It could hardly be expected that Crowe would freely offer the benefit of his personal observations to members of the financial community in both Canada and the United States, and at the same time withhold such views from government officials in Ottawa. Nor did he. He expressed his views in personal conversations to government officials, at least as high as the deputy minister level. Members of the bureaucracy involved in one way or another with the pipeline question, and with whom I later spoke in Ottawa, were all aware of Crowe's views. "There was no question in the minds of anyone in the bureaucracy as to where Crowe stood on the matter," one of them told me. "His views were widely known."

Late in 1977, I interviewed Crowe at his office in the Trebla building. I asked him if he was not concerned that his presentations at the two luncheons sponsored by Foothills financial advisory firms might entail risks of further controversy similar to that resulting from his tour to the Arctic Islands. He responded that he had in no way acted improperly in any of these events, that they involved people with whom he must associate in the discharge of his responsibilities, and that they were perfectly normal contacts. He said that in his addresses to the two meetings he had talked about "the whole scope of energy matters," with only passing reference to the applications, confined to his own personal observations, and giving as objectively as possible the pros and cons of all proposals. He said that since he was not involved in any way in the hearings, nor the board's decision, he was not restrained from expressing personal views.

Crowe also confirmed to me that he had been a dinner guest at Blair's ranch near Calgary at the time that the pipeline hearings were in progress before the board.

Marshall Crowe resigned from the National Energy Board at the end of 1977, having completed four and a half years of his seven-year term of office. He was later appointed chairman of a new government relations consulting firm in Ottawa, PIR Public Affairs Ltd., affiliated with Public & Industrial Relations Limited of Toronto. Among the clients of this new government firm is Dominion Securities Corporation, Harris & Partners Limited, another of the seven Canadian and American firms which have acted as financial advisors for the Foothills Alaska Highway project. Crowe also joined the board of directors of Energy Ventures Inc., a Boston firm affiliated with Loeb Rhoades & Co.

Crowe was succeeded as chairman of the NEB by Jack Stabback, who chaired the board's northern pipeline hearings, and whose entire professional career has been in the regulatory field, first with the Alberta Energy Conservation Board and then with the NEB.

I talked also to Stabback, prior to his appointment as NEB chairman, and he described the measures employed by the hearing panel to ensure that the board's decision was kept confidential prior to the announcement on July 4:

> We of course try to keep confidentiality on any case that we are working on. All draft material in the preparation of the report is marked confidential and the staff are under instructions to keep all working papers confidential ... When a panel of the board is struck to hold a hearing, that in effect becomes the board for the purposes of that decision. It is not subject to further review by either the board or the government, except that it is subject to the approval of the Governor-in-Council. So the panel of the board and the staff with which it works have to keep these materials confidential not only from other parts of the board, but naturally beyond the board to other parts of the government ... Because it was a high profile hearing, we obviously checked our security arrangements just to make sure that there wasn't a possibility of a leak.

Stabback explained that the "evidentiary" portions of the report were written by staff members, but that the decision itself was written by the three-member hearing panel. The staff, in writing the evidentiary portion, were kept uninformed of what the board's decision was.

I asked Stabback if political considerations influenced the board's decision. The hearing panel, he said, "was not insensitive to evidence we were hearing related to the concerns in the north. Now if that's political, small p, yes, that's true ... The environmental and socio-economic evidence that we got clearly indicated that this was a pipeline different from other pipelines that the board had been involved in. It was something that we definitely did feel quite strongly about and that we felt had to be taken into account." Had the board considered any political circumstances other than just that in the evidence before it? "We had no contact at all with departments or ministers in Ottawa," Stabback said. "We were completely cocooned. We had to be very sensitive to the evidence that was before us. Obviously you get impressions not only from the evidence but from everything else that you hear around, that there were concerns that people in general in Canada seemed to be having to a much greater extent than ever expressed before." Did the things which the board heard include the views of Marshall Crowe? "No. There was no inclination on his part to do so, and there was no inclination on the part of the panel to seek his advice."

In Ottawa, I found not the slightest shred of evidence that the hearing panel's decision was reached in anything other than the cocoon that Jack Stabback described. This does not mean, however, that in this cocoon the board did not consider the political implications. In fact, it could hardly ignore them. The panel would have had to be far more isolated to be unaware that, in light of Berger's report, approval of an Alaskan North Slope and Mackenzie Valley pipeline route would have confronted the government with difficult political problems.

Whatever the decision, there were serious political implications which demanded consideration. If the board did not consider them, then no one did, because it was the board, not the government, which made the decision.

The hearing panel's cocoon, that is, was not isolated from the political climate. That climate was generated largely by the Berger Inquiry, but there were other contributing causes as well, such as the split decision of the FPC. The pressures on the hearing panel to conform to the prevailing climate would be great. No one would need to tell them that the board's public and political credibility, already under question, would be subject to further strong attack from their critics in the environmental and so-called public interest organizations, as well as from the NDP, if their decision ran counter to Berger's findings and other prevailing currents. The unresolved question is, to what extent did Crowe's "personal observations" contribute to the prevailing climate in which the board had to make its decision? His views were known to the investment community in Canada and in the United States, and to the bureaucracy in Ottawa. Even in Washington, the FPC commissioners could hardly have been completely unaware of the freely expressed views of the chairman of the National Energy Board. The idea that the case for Arctic Gas was weakened by disappointing exploration results in the Delta, expressed by Crowe to the Toronto meeting April 26, was a view that greatly influenced the FPC decision, even though it was not shared by either the presiding hearing examiner nor the FPC staff.

The National Energy Board has the authority to reject any application for leave to construct a pipeline, without reference to the government. The government can reverse such a board decision only by having legislation passed in Parliament. But the board cannot independently approve a pipeline application, in effect it can only recommend approval to the government.

The board's rejection of one of the two applications, thereby reducing the options available to the government, plainly upset Energy Minister Alastair Gillespie. "I was surprised by the NEB report," Gillespie told me when I interviewed him several months after the board's decision:

> I expected they were going to come down with both options...
> I expected them to say that under certain conditions we would certificate the Mackenzie Valley route and under certain conditions we would certificate the Alaska Highway route. That would have left the government with a lever, so that in the negotiations with the Americans we'd have something to play. I like to think of it to some extent, in the final stages, as a horse race in which there were three main horses. There was one American horse [the El Paso proposal] and there were two Canadian horses. And they were all running very strongly. Then, one of the Canadian horses got shot, just before the finish line. I think it quite frankly removed some of the Canadian negotiating leverage. We would have been in a stronger position negotiating with the Americans if the board had not shot the Arctic Gas horse.

Right now the government is stuck with the Alaska Highway and Dempster routes, and Gillespie, for one, appeared none too sanguine about the prospects of these two systems securing the required financing.

"When it comes to financing, I've got to say that I'm a conservative," Gillespie told me. "Until the thing is all signed up, I'm not optimistic. I think the financing has got to be tricky...it's not assured. The government has received assurances from the Foothills – Yukon group that it will not require government financing or government guarantees, but until the whole thing is put together, I'm really keeping my counsel."

Does he anticipate a request to the Government of Canada for financial help? "In a theoretical sense I think you have to anticipate it," Gillespie said. "But for practical purposes, we've said no. It's been put to us on the basis that it's not necessary. These assurances have been given to us by the company, and we don't intend to guarantee it."

Neither do the pipeline companies, nor the United States Government, nor the State of Alaska, nor the oil companies. And if no one guarantees the pipelines, they won't be built. That is a prospect which must cause some concern in the Trebla Building.

Chapter 9
Environmentalists and the Cause Célèbre

In the plans for a pipeline along the Arctic coastal plain of Alaska and the Yukon, the environmental organizations found yet another cause célèbre. Like others, this particular cause célèbre involved great environmental problems standing in the way of public need, without dealing with the present realities of greater environmental loss. The public need is for the energy resources of the Arctic coastal plain. The perceived risk is principally the possible effect of the pipeline on a large population of caribou. The ignored reality is the indiscriminate shooting of caribou which has been and still is decimating their numbers.

The Arctic coastal plain, stretching more than seven hundred miles from Point Barrow eastward to the Mackenzie River Delta, may be the richest oil and gas region in North America. The plain spreads inland for a distance of up to one hundred miles, narrowing at the Alaska-Yukon border where it is pinched by the British Mountains and the Arctic Ocean to a distance of some ten miles. Prudhoe Bay, in the western section, is the largest oil and gas field on the continent. Geologists believe that many more oil and gas fields may be found here, possibly some as large as Prudhoe Bay.

The coastal plain fringes the Arctic wilderness. It is the brief seasonal haunt of migratory wildlife, and skirts the calving grounds of caribou. It was the key environmental factor in the decision to build two pipelines, rather than one, to transport the natural gas discovered at Prudhoe Bay and the Delta. Without a connecting pipeline across the coastal plain, there is no practical alternative to the construction of separate pipelines from each of these gas supply areas.

Construction of this four hundred miles of pipeline from Prudhoe Bay to the Delta would, it was deemed, have more adverse environmental effects than construction of an alternative fourteen hundred miles of northern pipeline along the Alaska and Dempster highways. That, at least, was the premise on which both Judge Berger and the National Energy Board concluded that an Alaskan North Slope–Mackenzie Valley pipeline is environmentally unacceptable.

A pipeline across the coastal plain was rejected because it traverses what was said to be a pristine wilderness and presents what were considered to be unacceptable risks to the Porcupine caribou herd whose summer calving grounds lie adjacent to this route. The Alaska Highway and Dempster routes (the latter cutting across the Peel River Wildlife Preserve as well as the wintering range and the spring and fall migration routes of the Porcupine caribou herd) were found to be more environmentally acceptable because they follow already established transportation corridors where the environmental effects of a pipeline, it was said, will be only incremental.

There are two views on this assessment, and both have been argued by regulatory authorities, as well as by northern environmental experts. Regardless of which of these assessments may be correct, there are four relevant facts which are not in dispute by anyone.

1) The longer Alaska Highway and Dempster routes are far less preferable in energy conservation terms, requiring much greater volumes of energy, as well as other resources, to transport the gas than the shorter route along the Arctic coastal plain.

2) Construction of the Dempster Highway from Dawson City to Inuvik involves much greater risks to caribou and other wildlife than would a pipeline along the Arctic coastal plain.

3) Far less is known of what the environmental effects really will be of the Alaska and Dempster Highway routes, because only the most cursory environmental studies have been made so far.

4) The worst of all circumstances, from an environmental viewpoint, would be the construction of pipelines along all three routes: the Alaska Highway, the Dempster, and along the Arctic coastal plain. Yet the Alaska Highway and Dempster pipelines do not exclude the possibility (some have said the inevitability) of later pipeline construction across most, if not all, of the Arctic coastal plain.

130

Justice Berger and the National Energy Board were both emphatic in their view that construction of the four hundred miles along the coastal plain (of which more than half would be in Alaska) involves environmental risks which are unacceptable. According to Justice Berger:

Gas pipeline and corridor development along the coastal plain passing through the restricted calving range of the Porcupine caribou herd, would have highly adverse effects on the animals during the critical calving and post-calving phases of their life cycle. The preservation of the herd is incompatible with the building of a gas pipeline and the establishment of an energy corridor through its calving grounds. If a pipeline is built along the coastal plain, there will be serious losses to the herd. With the establishment of the corridor I foresee that, within our lifetime, this herd will be reduced to a remnant. Similarly, some of the large populations of migratory waterfowl and sea birds along the coastal route, particularly the fall staging snow geese, would likely decline in the face of pipeline and corridor development.

The National Energy Board was just as conclusive in its findings. It ruled that the coastal route is "environmentally unacceptable," because it involves "impacts which could not be avoided, which could not be accepted, and for which mitigative measures are unknown or uncertain of development." The board found that the evidence of Arctic Gas on the coastal route "is not compelling enough to conclude that the construction, operation and maintenance activities for the proposed pipeline in the northern Yukon would have little or practically no adverse effect upon the Porcupine caribou herd, particularly in the event of any pipeline-related emergencies or of emergency repairs to the pipeline during the critical periods of the herd's calving and post-calving aggregation."

Even limiting actual pipeline construction to the winter months, when the caribou are several hundred miles away in their winter area in the Porcupine Hills, would not avert the risks, because there would still be substantial related activities in the summer, according to the board. It noted that activities "such as construction of wharf sites, roads, airstrips, stockpile sites and compressor stations, excavation and stockpiling of granular material from flood plains, rock blasting and quarrying, transport of heavy modular units from

coastal wharf sites to compressor station pads, aircraft and helicop-
ter movements . . . would all take place in spring or summer. The
board is not convinced that the impact of these activities on the
caribou herd would be nil, or at worst minimal," as claimed by
Arctic Gas.

Neither did the board accept the position of Arctic Gas that it
would be able to time actual construction work in this area to
periods when the caribou are absent, since it found that the migratory patterns of the herd show great variation. Thus it concluded
that "avoidance of the herd is highly uncertain" and that "the
applicant's statement that the impact of the pipeline on the caribou
herd would be minimal," is too optimistic.

Far less uncertain, however, is whether construction of a pipeline
along the route of the Dempster Highway could be timed to avoid
contact with the herd. It can't. The herd spends about three quarters of a year in the region that would be traversed by a Dempster
pipeline, and only a comparatively short time on the coastal plain.

Not all of the assessments of the impacts of a pipeline on the
wilderness values of the Arctic coastal plain and on the caribou
were as negative as those of Justice Berger and the National Energy
Board. More than half this route lies in Alaska, and the U.S.
Department of the Interior, the staff of the Federal Power Commission, Judge Litt and FPC Commissioners, all found this route to be
environmentally acceptable, if not preferable.

Donald H. MacKay, professor of chemical engineering and associate director of the Institute for Environmental Studies at the
University of Toronto, believes the route along the Arctic coastal
plain and the Mackenzie Valley to be environmentally preferable.
His statement, presented to the Mackenzie Valley Pipeline Inquiry
in Toronto, marked one of the few occasions when Judge Berger,
who listened with impassive interest during eighteen months of
public hearings, registered notable surprise.

"I am convinced that by the early to mid-1980's Canada could
face a severe petroleum energy shortage," Professor MacKay told
the judge. "From the standpoint of energy needs alone, it is impera-
tive that exploration proceed as fast as possible in the Mackenzie
Delta and Beaufort Sea and that gas and oil pipelines be constructed at the earliest possible date. The longer the delay in prov-
ing reserves or constructing viable transportation systems, the more
critical the national energy situation may become and the more

likely it is that environmental and social factors will be disregarded. I'm thus in favour of early construction of the Canadian Arctic Gas pipeline."

"Did I hear you to say that you favoured specifically the Arctic Gas pipeline?" the judge interjected.

"Yes, your honour; the Arctic Gas pipeline," MacKay responded. He continued: "In the long-term, we must depend on renewable energy resources, but there is simply no possibility of substantially decreasing our dependence on oil and gas in the next ten to fifteen years. The industrial and social consequences of a short-fall in oil and gas supply could be disastrous. I am convinced that it would be intolerable to proceed with these developments in an environmentally and socially damaging manner."

Professor Lawrence Bliss of the Department of Botany at the University of Alberta, an ecologist who has spent more than twenty years of research and teaching on Arctic and alpine environments, asserts bluntly that the combined environmental effects of the Alaska Highway and Dempster pipelines will be "far greater" than the shorter, single pipeline route.

"As an ecologist I favour the coastal land route from Prudhoe Bay and not the unstudied Alcan [Alaska Highway] pipeline route for the reasons that gas will be brought south from the Delta and more gas will be found offshore in the Beaufort Sea westward," Dr. Bliss told a seminar at the University of Toronto shortly after the release of the Berger report. "The holistic approach of a single northern gathering system bringing gas from the known and highly probable new fields in one pipeline is the most sound ecologically, environmentally, economically and probably socially. A holistic approach to current and potential future pipeline plans is needed now."

Bliss has been involved in basic and applied research in relation to petroleum development in the Mackenzie Delta and has served on both governmental and industry sponsored committees "to help determine how best to conduct northern development." He later expressed his views in a letter to Prime Minister Trudeau. He claimed that a pipeline across the coastal plain "will do far less environmental and ecological damage than two or three pipelines."

Bliss was also upset because the Berger report paid scant regard to the findings of professional environmental researchers who had spent five years examining the environmental impacts of a pipeline along this route and developing mitigative measures. He was a member of the Environmental Protection Board, an independent body of ecologists and engineers who have performed studies funded by Alberta Gas Trunk Line, Arctic Gas and Foothills Pipe Lines.

The EPB was involved in some of the twenty million dollars in environmental studies conducted for Arctic Gas. "Never before in North America has industry spent so much on ecological research and then openly published the findings in a widely-acclaimed forty-one volume biological report series which was made available to universities throughout Canada," according to Bliss. (Judge Litt noted that this "superb" series of reports "has significantly advanced the state of scientific knowledge".) Bliss complained that, "the Berger report gives little credit to the studies, sometimes even to the scientists who supervised or evaluated the work." He is concerned that this could impair future opportunities for ecologists to work on comprehensive studies funded by industry. To ignore these studies and approve the highway routes where studies have not been conducted "would make a mockery of the process we have all been involved in over the past six years," Bliss wrote to the Prime Minister.

"There can be no question that of all mammals [along the North Slope of Alaska and the Yukon], the Porcupine caribou herd represents the single most important consideration," Judge Litt found in his report, a view reinforced by Berger and the National Energy Board.

This herd numbers up to an estimated one hundred and twenty thousand animals who have not yet found out that there is a border between Alaska and the Yukon. They wander freely back and forth from one territory to the other as if the border did not exist, truly citizens of both countries.

However, they are, according to Dr. Frank Banfield, perhaps the world's foremost authority in caribou, "stolid" animals, phlegmatic creatures who are not easily spooked. In Yellowknife, I spoke to a bush pilot who considers caribou to be a "bloody nuisance." He told me about the problems they caused on one occasion when he wanted to land a Twin Otter at Coppermine. "They were all over the damn strip," he said. "I buzzed them once, and they wouldn't budge. I tried again, and they just looked at me with their stupid faces. The people of the village finally had to come out and chase them away before I could land."

In Old Crow, I was told that the caribou have been known to walk right through the village on their annual northern migration. A group of caribou once walked through the middle of a camp site set up by a field party of biologists who were attempting to trace their migration patterns.

I once watched a small group of caribou pass by an electronic device which had been set up to simulate the noise of a pipeline compressor station powered by aircraft-type jet turbines. The device sounded like a DC8, a high frequency noise measured at ninety-four decibels from a distance of fifty feet. It was late April at Chute Pass in the Richardson Mountains, seventy miles west of Old Crow, and the snow was still ass deep. We had struggled on snowshoes half a mile up the side of a mountain to a lookout site built of snow blocks. There were Ron Jakimchuk and Dean Feist of Renewable Resources, an Edmonton environmental consulting firm; Don Thomas, an environmental writer with the *Calgary Herald*, and Fred Frost, a native assistant from Old Crow. (Feist was later killed when an aircraft from which he was spotting caribou crashed into the side of a mountain.) Below us, in the narrow mountain pass, we could see the caribou walk past within three hundred feet of the sound simulator, seemingly unperturbed by the noise.

Judge Litt, on an "official view," a tour of the proposed pipeline routes, in August, 1976, also found the caribou to be stolid creatures:

> ... the view not only gave an impression of the physical aspects of the area to be traversed and the sites, but observation of the area also led to impressions as to the weight to be afforded some of the evidence which, on its face, was inconsistent with what was seen. Caribou grazing on fields surrounded by gravel roads, pipes carrying oil to the pump stations, and oil field construction and industrial facilities give a different impression of the compatibility of some caribou with industrial areas than the record might have indicated. Similarly, an eagle's nest with fledgling birds just a few feet from the Alyeska main road to the Valdez oil terminal under construction gives a strong impression of at least one set of eagles' sensitivity to man's activities.

Litt reported that "one of the caribou observed from the tour bus at Prudhoe Bay clambered onto the road a few feet in front of the bus when the bus stopped, crossed in front of the bus, and went to another field on the other side of the gravel road."

The most critical period in the life-cycle of the Porcupine caribou herd is during the calving and post-calving aggregation period. This occurs between late May and mid-June, over an eight thousand square mile area between the Babbage River in the Yukon and the Katakuruk River in Alaska. Most of the calving takes place in the foothills rather than on the coastal plain. It is at this time and place that the caribou will be most susceptible to disturbance by men, but exactly how susceptible is uncertain. Judge Litt concluded that there is no "significant evidence that the plans of any of the pipeline applicants will adversely disturb, in the short or long term, any of the caribou along those pipeline routes." Dr. Bliss is critical of at least one aspect of the environmental studies, claiming that biologists did not spend enough time testing to see how caribou reacted to noise during the calving period.

In his paper for the seminar of ecologists at the University of Toronto in 1977, Bliss wrote that, "unfortunately the caribou researchers spent much time and money counting animals, determining their patterns of annual migration, studying their behavior, but unfortunately did little experimental research to determine the quantitative impact of low flying planes on caribou at the time of calving and post-calving aggregations. The limited data... showed that cows are protective of their calves and only under severe duress did the cows run off. Everyone agrees the calving period is critical, but until biologists are willing to conduct actual destructive experiments to determine the levels of impact, they are in no position to enter into the critical decision process expected of them by government and industry. Thus an important issue... was not adequately answered."

Nevertheless, what limited evidence is available, appears encouraging. Frequent aircraft flights over and into the area during the summer months have been underway for several years, and Bliss claims that "no researchers have reported any serious impact from these current activities." He notes that a research report "by three men who studied the herd shows that on two occasions cows with new-born calves paid no attention to several low passes of the aircraft, even at altitudes of one to two hundred feet. On other occasions cows stayed with dead or sick calves even with helicopters nearby. Although the data are limited, they clearly show that cows

136

and new calves are no more sensitive to aircraft disturbances than they are at any other seasons." Bliss asserts that "to state, let alone imply, that pipeline activity in summer could or would severely impact on this herd of very important animals is not supported by the facts."

The cause célèbre of the environmentalists comes alive at the mention of Alaska's Arctic National Wildlife Range and its mystique of pristine wilderness that must not, it is said, be violated by man. The philosophical arguments of the conservationists, notes Litt, "are couched in absolutes which appear tautological; any construction will defile the virgin territory and the virgin territory is defiled by any construction."

The Arctic Wildlife Range is neither a wilderness area legally under the terms of the American Wilderness Act, nor, at least along the coastal plain, is it a true wilderness area in fact.

Established by a Public Land Order issued in 1960 by the U.S. Secretary of the Interior, the Wildlife Range covers a total area of fourteen thousand square miles (five thousand of plain and nine thousand of foothills and mountains). Its total area is more than six times the size of Prince Edward Island, and it is bounded on the east by the Yukon border and on the north by the Arctic Ocean.

This range, particularly the coastal plain, is not exactly virgin wilderness. At the turn of the century, whalers completely wiped out the entire population of musk oxen in the area. Today, the landscape is marked by a rusting hulk in Demarcation Bay, numerous temporary native hunting and fishing villages and two inactive Dew Line stations. In addition, "Smack in the middle of the coastal section of the Wildlife Range is an operating Dew Line site with multisensory radar receptors, an airfield, docks, boats," Judge Litt notes in his report. "Next to it is Kattovik [a native village with a population of two hundred]. Nothing appears to rust very fast in the Arctic, and because of the permafrost, nothing, or little is buried. And, since the terrain is flat, the debris of civilization, broken snowmobiles and beached boats, discarded appliances, etc. are visible around and about the houses of the native villages." The actual, physical impact that a pipeline would add to this picture would be contained within four square miles out of fourteen thousand, most of which would be the right-of-way across the buried pipeline and comparatively little devoted to ancillary facilities (compressor station, air strip and staging depot).

Other petroleum development activities may well intrude upon the coastal section of the range, even though a pipeline to connect the Prudhoe Bay and Delta gas reserves how now been ruled out.

Because it is a Wildlife Range and not a designated wilderness area, the Interior Department is at liberty to permit controlled resource development on the range, providing that such activity is not incompatible with wildlife resources and recreational uses. "The language on the face of the Public Land Order [establishing the range] specifically provides for issuance of permits under the Mineral Leasing Act," notes Litt, "a totally useless verbiage if a pipeline right-of-way is excluded." In addition, the Refuge Administration Act permits the Secretary of the Interior to:

> ... permit the use of, or grant easements in, over, across, upon, through or under any areas within the System for purposes such as but not necessarily limited to, powerlines, telephone lines, canals, ditches, pipelines, and roads, including the construction, operation, and maintenance thereof, whenever he determines that such uses are compatible with the purposes for which these areas are established.

The economic pressures on the United States to open the five thousand square miles of the coastal plain within the Arctic Wildlife Range to petroleum exploration may sooner or later become overwhelming.

The remaining discovered oil reserves in the United States in 1978 amounted to some thirty billion barrels. The Prudhoe Bay field accounts for one-third of this total. Even larger reserves might be locked beneath the coastal plain of the Arctic National Wildlife Reserve.

Some indication of the oil potential of the Wildlife Range is contained in a 1974 report of the U.S. Department of the Interior (*Final Environmental Statement, Proposed Arctic National Wildlife Refuge*). The reservoir rocks which contain the oil at Prudhoe Bay probably underlie the entire five thousand square miles of coastal plain in the range, providing room, according to the Interior report, "for several structural traps similar to those at Prudhoe Bay ... In addition, the Wildlife Range has potential reservoir rocks both older and younger than those at Prudhoe Bay ... The presence of oil in some of these sediments is demonstrated by seeps along the coast at Barter Island and Angun Point and oil saturated outcroppings on the lower Katakturuk and Jago Rivers." The most striking prospect is known as the Marsh Creek anticline, described by the Interior Department as a "potential oil bearing structure" measuring forty-six miles in length and covering one hundred and fifty thousand acres. "At least four geological formations may harbor oil and gas" in this structure, with "potential for a six billion barrel oil reserve." The State of Alaska Division of Geologic Survey has indicated that a twenty billion barrel potential reserve—two-thirds of the present total U.S. oil supply—is entirely reasonable.

Until exploratory wells are drilled, no one will know whether the Wildlife Range does, in fact, contain oil reserves of this magnitude, or whether it contains any oil at all. But with so much at stake it would seem that the United States can hardly afford not to find out. Judge Litt considers it very likely that such exploration will take place: "It is clearly in the public interest to exploit hydrocarbon reserves," Litt reported, "and unless Congress unequivocally prohibits such exploitation in or off the Wildlife Range, the ultimate incursion into the range for such exploitation must be considered to be a virtual certainty."

If such incursion does take place, the decision to build fourteen hundred miles of gas pipeline in order to avoid four hundred miles of line across the Arctic coastal plain will prove to be the ultimate folly.

Judge Berger, however, has urged the United States to accord wilderness status to the Wildlife Range and legislation to accomplish just that has been proposed by the Carter Administration.

In the first volume of his report, released in May, 1977, Berger wrote that if an Alaska Highway pipeline route were approved, then "any agreement in this regard between Canada and the United States should include provisions to protect the Porcupine caribou herd and the wilderness of the northern Yukon and north-eastern Alaska. By this agreement, Canada should undertake to establish a wilderness park in the northern Yukon, and the United States should agree to accord wilderness status to its Arctic National Wildlife Range, thus creating a unique international wilderness park in the Arctic."

Berger again stressed this proposal in the second volume of his report, released early in 1978. "With the rejection of the Arctic Gas pipeline proposal, there is now an opportunity to plan for land use in the Mackenzie Valley and western Arctic without the pressure of imminent, largescale industrial development. Wilderness areas, if they are to be preserved, must be withdrawn from any form of industrial development," Berger urges. "That principle must not be compromised."

Apparently, Berger feels that the development of the petroleum resources of the range will benefit only industry. "Conservation areas should not be selected only from those lands that are of no value to industry," he states. There is no recognition of any public interest in acquiring energy; it is merely the interests of industry that stand in the way of wilderness values.

Less than five months after the first volume of Berger's report was issued, the Carter Administration proposed amendments to the Alaska National Interest Lands Conservation Act. The amendments would increase public conservation lands in Alaska from thirty million to one hundred and twenty million acres, or one-third of the state. Of the existing and proposed conservation lands, wilderness status would be accorded to forty-three million acres, including the Arctic National Wildlife Range. This would automatically preclude any oil and gas development.

There are strong Congressional pressures to exempt at least the five thousand square miles of coastal plain in the Wildlife Range from wilderness status because of the possibly enormous energy resources that the United States desperately needs. Trade-offs have already been suggested involving wilderness areas that are more truly wilderness in their nature.

Judge Berger has pressed his case to ban all industrial activity along the coastal plain not only in his two reports, but also in American speaking engagements and before U.S. Congressional Committee hearings early in 1978. But despite Berger's pleas, it cannot be assumed that the United States will blithely ignore one of its best hopes to significantly increase its oil and gas supplies.

This prospect is not the only aspect which appears to be over-looked in the environmental assessment of alternative pipeline routes for western Arctic gas, because neither Judge Berger nor the National Energy Board appear to have really taken conservation into account. The environmental and conservation aspects of energy development are as inseperable as bread and butter. The most frequent argument of environmentalists opposed to any particular energy development is that it would be unnecessary if we could only learn to curb the voracious energy appetite of our industrial and materialistic society. Judge Berger refers to "the economic religion of our time, the belief in an ever-expanding cycle of growth and consumption," and asks: "Will we continue, driven by technology and egregious patterns of consumption, to deplete our energy resources?" Thus it is difficult not to consider both the environment and conservation when thinking about energy.

Yet, Justice Berger and the Energy Board both failed to consider energy conservation in finding a pipeline across the Arctic coastal plain to be environmentally unacceptable. In its decision of July 4, the board espoused "a vigorous conservation policy," but nowhere in its report does it consider the energy that would be lost in transporting gas from the western Arctic through an additional one thousand miles of pipeline.

Gas pipelines use a portion of the gas that they transport to fuel the compressor stations. The Alaska Highway and Dempster pipelines would use in the order of seventy-five per cent more of the gas in the Delta and Prudhoe Bay than a single, shorter pipeline along the Arctic coastal plain and the Mackenzie Valley. The two pipelines would burn up close to an extra one trillion cubic feet of gas. The additional gas consumed would be enough to heat three hundred thousand Canadian homes for a period of twenty years.

The essential features of an environmental cause célèbre— whether it be the seal hunt off Newfoundland or the Storm King electric power development on the Hudson River—seldom vary. They are: a dramatic circumstance with strong emotional appeal which provides the advocates with the uplifting sense of a fashionable cause, where perceptions are everything and realities are irrelevant.

Perhaps the classic example of the price of this type of environmentalism is the controversy around the proposed "pump-storage" hydro-electric project at the base of Storm King Mountain on the Hudson River, some one hundred miles upstream from New York City. The story of the Storm King battle, chronicled by William Tucker in the December, 1977 issue of *Harper's* magazine ("Environmentalism and the Leisure Class: Protecting Birds, Fish and, Above All, Social Privilege") contains some interesting parallels for the environmental cause célèbre on the Arctic coast. First proposed in 1962 by Consolidated Edison to provide some urgently needed peak electric power at an estimated cost of one hundred and fifteen million dollars, Storm King Mountain has been stalled for fifteen years by demands for repeated reviews by the Federal Power Commission of "new evidence" and by endless court litigation. During these fifteen years, escalation and revised plans to accommodate environmental objections have increased the cost of the Storm King project to one billion dollars; New York has suffered a series of devastating power failures; Con Edison was pushed to the brink of bankruptcy; industry fled New York with a resultant loss of six hundred and fifty thousand jobs, in part because of electric power shortages and soaring rates.

One argument advanced against the project by the Scenic Hudson Preservation Conference was that "the Atlantic striped-bass population which spawns in the Hudson will be destroyed for all time." A four hundred and fifty thousand dollar study conducted by Con Ed and supervised by state and federal authorities had estimated that the plan might result in a three point six per cent mortality rate of all fish eggs and larvae during the May-July spawning season. "Since 99.99 per cent of all eggs and larvae die before maturity under natural conditions, the conclusion was that the plant would have no effect on the total fish population," Tucker writes. Scenic Hudson made monumental errors in interpreting the study data—whether deliberately or not—counting the annual mortality of eggs and larvae as the daily mortality and the loss of eggs and larvae as loss of fish, thereby claiming the project would kill seventy-five per cent of the hatch in the seven-week spawning season. Senator Edward Kennedy was moved to plead with the FPC for a further review because of this impending disaster. The further review was a twenty million dollar, four-year study involving seven major universities, all of which confirmed the earlier study, estimating that the plant would destroy four to five per cent of the eggs and larvae and have no effect on the total fish population.

While decrying this mythical threat to the Atlantic striped-bass, the Scenic Hudson Preservation Conference turned its blind eye to the real loss resulting from excessive sport fishing. "There was a certain irony here," writes Tucker, "since sports fishermen had long since destroyed major stocks in almost every river and stream in the Northeast, and the federal and state governments were annually spending hundreds of thousands of dollars for restocking so that they could continue their pastime."

The driving force behind the environmentalist movement, concludes Tucker, is the affluent leisure class which stands athwart any progress that threatens the status quo: "members of the local aristocracy, often living at the end of long, winding country roads." People who already have it made under the status quo are apt to instinctively find any change in the status quo a threat to their interests. "Environmentalism," writes Tucker, "always seemed to work in favor of the people who were already established in 'the environment'." Wealthy urbanites with country estates want no power lines to blot their views; members of exclusive resorts fear any incursions which may touch their sanctuaries; hunters who fly in private aircraft to remote, inaccessible regions want those regions left just as they are. "The environmental vision is an aristocratic one," writes Tucker, "at the point where an idyllic past blends nicely with an imaginary future. It can only be sustained by people who have never had to worry much about their security."

Tucker argues that the grave risk in opposing projects needed now (like new oil and gas supplies) for options that are distant or illusory (solar collectors basking in the sun and windmills gracefully spinning in the breeze) is that society may be driven to extremes of permanent wealth and permanent poverty. Both will oppose change: the wealthy because they don't want it, and the poor because they are not equipped to cope with it. The vision of environmentalism, says Tucker is "a vision that will call us to disaster."

There are striking similarities in the efforts to save Storm King Mountain and the efforts to stop any petroleum development in the Arctic National Wildlife Range. Many of the organizations leading both battles were the same: the Sierra Club, the Wilderness Society, the Audubon Society and others.

In both cases, reality became distorted. On the Hudson River, the decaying industrial area near Storm King Mountain became a scenic asset. In the Arctic, the Wildlife Range, with its abandoned ships, Dew Line sites, air strips and native villages, became a virgin wilderness, the "crown jewel of the Arctic." In both cases, the environmentalist organizations managed to influence the outcome of quasijudicial proceedings more by public perception than on the basis of the hearing records. "Court decisions aren't made in a vacuum," a spokesman for Scenic Hudson is quoted as stating, "Judges read newspapers."

In both cases, the small valiant band of environmental groups is seen as fighting for the public interest, pitted against the awesome corporate power of big and resented firms. In the Arctic, it is the big oil companies. In New York it is giant Consolidated Edison. "It's really the public interest versus the big brother or big daddy approach of Con Edison, which assumes we should let it tell us what is good for the public," said Scenic Hudson as it pleaded for its wealthy members with their rural estates. That claim sounds hauntingly similar to many of the presentations by special interest groups to the pipeline hearings. One spokesman for Scenic Hudson reportedly went so far as to suggest they were out to financially ruin Consolidated Edison. "We wanted them to go into receivership so that even we could buy them out—then they would be a true public utility."

But the most striking common feature is the manner in which, in both cases, the environmental groups turned a blind eye to the real danger to wildlife. In the Hudson, it is excessive sports fishing of the Atlantic striped-bass. In the Arctic, it is depletion of the caribou by excessive hunting.

It is not, according to Judge Litt, the plan for any pipeline construction which "represents the significant factor in caribou herd viability in Alaska." It is hunting that poses the danger. "Man has been systemically destroying these animals through overhunting, whether by subsistence hunting or sport, to the point where the herds may not be able to maintain population levels necessary for survival." The Forty Mile herd of caribou in Alaska, estimated at five hundred thousand at the turn of the century, had been reduced by hunting to eight thousand by 1977. Litt warned: "The same could well happen to the Porcupine herd, if the same hunting laws are in force, and for the state and federal officials to sanctimoniously enter into discourses as to whether a pregnant caribou will or will not cross a one or two-foot high berm, while permitting almost indiscriminate shooting of these animals by any Alaskan along the migratory route, or any rich hunter wanting a 'double shovel' set of antlers, is illogical."

Nary a word of this threat to the caribou herd was whispered by the environmental organizations during the pipeline debate. The reason is not difficult to discern. The subsistence hunting of caribou by native people is not the proper material for a cause célèbre. Neither is shooting by wealthy hunters who fly into the remote wilderness in private aircraft, armed with high-powered rifles and telescopic sights. These are sportsmen, people who champion protection of the priceless heritage of wildlife, and donate fat cheques to support the environmental and wildlife organizations.

But oil wells, gas wells, pipelines, and the Arctic National Wildlife Range—now there is the material for a cause célèbre.

Chapter 10
The Pipeline, Political Theology and Other Assorted Marxists

It was once called "the Conservative Party at prayer." It represented The Establishment in Canada. Today, its ranks thinning, the Anglican Church of Canada represents but ten per cent of the total congregation of the major churches.

It still plays the role of the leader. Its head, the Most Reverend Archbishop Edward Walter Scott, Primate of the Anglican Church of Canada, is also the chairman of the central committee of the World Council of Churches, and has been described as "the most influential figure in Canadian religion." Its banner flutters in the vanguard of the ecumenical movement that unites the major churches, if not in one common body, at least in one common cause.

That leadership, in the Anglican and other major churches, is now taking the bands of diminishing and divided followers in new and controversial directions, leaving many behind at the latest sharp turn to the left. The parishioners—solid folk, hard-rock mainstream of society, acutely aware of the Biblical injunction that "by the sweat of thy brow shalt thou earn thy bread"—still gather in Sunday service to find inspiration and strength in the old familiar prayers and hymns. The thinkers in the churches gather in seminars to ponder the new "political theology," Bible in one hand and Marx in the other. And the doers in the churches gather, in the cause of social activism, to accuse corporations of wrongdoing at shareholders' meetings, to confront governments at public hearings with demands for social justice, and everywhere backing the liberation wars of the poor and oppressed, from Zanzibar to Aklavik.

No longer "the great defender," the Anglican Church now leads the others to challenge the entire established order, denouncing "a socioeconomic situation which is sinful," and endorsing the call for "a new international economic order."

It was all this, and more, which came together in the crucial role that the churches played in the pipeline debate. In retrospect, it can be seen that it was the churches, more than any other public group, which most influenced the pipeline debate and the final outcome. It was the churches who most actively worked with and helped finance the radical leadership of the Indian Brotherhood of the Northwest Territories. In public statements, in meetings with the Prime Minister and federal cabinet, in scores of presentations before the Berger Inquiry, in appearances before the National Energy Board, the churches stated their cause in rhetoric that was too often uncompromising, strident and filled with invective. They disputed the need for northern energy supplies; they raised the spectre of ecological disaster and devastation; they displayed an emotional xenophobia with strong inferences of domination by malignant American interests; and they accused both corporations and governments of deception and greedy motivations leading to purposeful exploitation and oppression. But most of all, in the name of a just settlement of native land claims, they demanded a moratorium on all northern development and construction of northern pipelines, a moratorium of at least ten years, or however long it might take to implement settlement.

In the end, the churches did not win the moratorium on northern development that they sought, just a moratorium on the construction of a pipeline along the Mackenzie Valley.

The valid concern of the churches with the gravely disadvantaged position of northern native peoples, and the need for a just and equitable settlement of their land claims as one of the essential means to help correct this, is beyond dispute. Equally valid is their concern for conservation and environmental protection.

What is at issue, though, even among many church adherents, are questions raised by the ways in which the churches pursued these objectives.

Do tactics of confrontation and vindictive accusations lead to productive results?

Do the churches have the competence—or even the right—to prescribe specific responses to highly complex social and economic matters which require detailed analysis and on which not all devout Christians agree?

Will a moratorium, in fact, help achieve just land claims settlements for the Dene, Inuit and Metis of the Northwest Territories?

Can these northern native peoples achieve social and economic justice without both an equitable settlement of their land claims and the provision of needed jobs through the development of non-renewable resources?

By demanding an open-ended moratorium, were church people (most of them unwittingly) playing into the hands of white activists within the NWT Indian Brotherhood whose primary objective was not the settlement of land claims but the establishment of a radical socialist state in the NWT?

Why was the demand for a pipeline moratorium, and support of the Dene nation concept, espoused by hundreds of church people from southern Canada, but not publicly endorsed by clergy from the north?

Does the antipathy to a profit-motivated market economy shown in many of the church submissions simply reflect concern about the system? Or does it also reflect Marxist elements within the churches?

I tried to find some answers to these questions. I examined many hundreds of pages of testimony by church people who testified before the Berger Inquiry and the National Energy Board, as well as other church documents and public statements. I spoke at length to several church people who were gracious enough to give generously of their time. What I found was widespread agreement that the churches have probably raised at least some of the right questions. But there is little agreement that they have found—or should even try to find—the right answers.

The two major churches, Anglican and Roman Catholic, have had a long and close association with the native peoples of the Mackenzie region, involving not only religion but also education, health and social services. First in the field were the Roman Catholics, who established the first mission in this vast area at Fort Providence in 1857.

Eric Watt, writing in the *Northern News Report*, June 9, 1977, portrayed some of the colour of the missionary work of the two churches in the Mackenzie:

> By the time the Anglicans realized there was a whole new land awaiting Christianity, the Catholics were firmly established as far north as Fort McPherson.

There was bitter inter-church strife when the Anglicans began to move in.... While that rivalry lasted well into the 1950's, when both churches found themselves suddenly assaulted ... by the Feds, who took over education, health care and many of the social services provided by the churches, the Hudson's Bay Company and the RCMP and other churches, a line of Anglican—RC demarcation was finally created at Aklavik.

By the mid-1950s, the Catholics claimed about 90 per cent of the Indians of the western NWT as their parishoners; the Anglicans, the same percentage of Inuit. Aklavik had both Anglican and RC residential schools, teaching up to junior high level. Those few native students who went past grade 8 had to go south for their schooling—to other church-run residential schools in the provinces.

... the missionaries of both faiths, regardless of who represented the minority, worked hard and faithfully for both their own congregations and the 'enemy's'. Many an Anglican Inuit at Holman, for instance, boasts fillings done by Father Tardy, the Oblate priest there, who had his own dentist's chair. And there are few priests of either faith who were not respected by the native people of those settlements.

The two mission schools at Aklavik were model schools in many respects. They became part of the community. Many a student came from places like Tuktoyaktuk or McPherson or Arctic Red River, which meant he or she was away from family life much of the year. But both schools had their own traplines, just out of the settlement, and students whose marks were good could go trapping every weekend during the season.

And those were happy hostels. Transients—the writer among them—frequently stayed there because there was no other accommodation.

The federal government's take-over of education, with the establishment of large, centralized hostels which took native children even farther from their families, "was an unmitigated disaster at its outset," according to Watt. "The churches dug in and bitterly fought the federal invasion, but they were too late. Civil servants now handle most of the welfare once looked after by the churches, the Bay and police. Government hospitals care for the sick. The churches and the missionaries, by and large, are back concentrating on church work, though still active in many fields of social assistance. And if the widespread ecumenical approach of the churches in the south isn't exactly copied in all northern communities, most Anglicans and RC missionaries have learned to live with, and often work with, each other and the newcomers."

According to Watt, the two churches that led the demand for a pipeline moratorium "apparently paid little or no attention to their own northern missionaries when their Toronto-based hierarchies jumped on the anti-development bandwaggon."

As early as 1972, Father Joseph Adam, a long-bearded patriarch who had lived some thirty-six years in Aklavik and Inuvik, warned that native people would "blow up the pipeline" unless it provided some opportunities for them.

"They are not opposed to it," Father Adam said on a CBC radio interview. "Oh no, I spoke to some of them and they are not opposed. In fact, they did not ask very much. But they wanted to have something to say in the pipeline, and especially jobs. They wanted jobs, too, for maintenance and things like that, as long as they are qualified of course...I am in favour of a pipeline or a railway, whichever best it is. I am certainly in favour of development of the north. It's got to be opened. But then, what I wanted to tell is that if the people are not satisfied, some will resort to violence."

Two years later, I had a luncheon meeting in Toronto with Right Reverend John R. Sperry, recently appointed Anglican Bishop of the Arctic. It is the largest diocese in the Anglican Communion, nearly three million square miles. A veteran of the Royal Navy during the Second World War, Dr. Sperry had served his ministry in the north for more than twenty years, at Coppermine, Fort Smith, Aklavik and Yellowknife. He speaks Inuit fluently, has translated a major part of the New Testament into Inuit dialect, and in the first ten years of his northern work had travelled nearly thirty thousand miles by dogsled. He had met with Mike Lewis and me to discuss some of his concerns about our proposed pipeline. We talked about these over lamb chops and apple cider in the somewhat faded elegance of the dining room at the King Edward Hotel. Later that day he pursued his concerns further in a chat with Bill Wilder.

Bishop Sperry's deepest concerns were clearly about the spiritual, social and economic well being of the native peoples of the north.

He wanted to know about the measures we proposed to protect the environment during pipeline construction and about our plans for training and employment of native peoples. He was very worried about the lack of employment opportunities for northern native peoples. And, to my surprise, he expressed grave concern that the land claims demands being formulated by the Indian Brotherhood were creating unrealistic expectations which he feared would result in frustrations and bitterness when final settlements failed to fulfill visions of affluent leisure. The bishop made it clear that in his view the employment opportunities afforded by pipeline-related development would be welcome in the north, providing that social and environmental concerns could be properly handled.

We tried to respond to the bishop's questions as fully as we could, and we invited him to contact us again any time he had further questions or wanted more information. That was the last that I, or, as far as I know, anyone else at Arctic Gas heard from Bishop Sperry.

The following January, after the public hearings before the Berger Inquiry had been announced, Bill Wilder wrote to Bishop Sperry, suggesting that the bishop consider presenting his point of view at the hearings. It was not, of course, an altogether altruistic suggestion, since we clearly felt that Bishop Sperry's viewpoint was supportive of our own to at least some degree. But we did think it was an honest suggestion, and that with his extensive first-hand experience and insight he was in a position to make a significant contribution to the hearings. Again, Bill Wilder suggested that should the bishop "require any further information, or have any questions, I hope that you will not hesitate to contact me." He never did receive a response to the letter, and Bishop Sperry did not testify before the Berger Inquiry, nor before the National Energy Board. It was the first indication that, while church people from Vancouver to Charlottetown said what they thought was best for the north, the clergy in the north were not included in the discussion.

HOW THE CHURCHES
HELPED SCUTTLE THE PIPELINE

There was a chain of related events and people that connected Church House, the headquarters of the Anglican Church of Canada at 600 Jarvis Street, Toronto, with the Yellowknife offices of the Indian Brotherhood of the Northwest Territories and the Berger Inquiry.

The first link in that chain may have been Peter Puxley, a young Rhodes scholar and son of an Anglican cleric, Canon H. L. Puxley, formerly principal of King's College, Halifax. Puxley, together with Stephen Iveson from North Bay, Ontario, arrived in Yellowknife in 1969 with the Company of Young Canadians, the federal government's discredited and later disbanded vehicle for youths to dedicate a part of their lives to help others, or, as it often turned out, to push a social or political cause at home or abroad. In Yellowknife, Puxley and Iveson worked with some articulate young native people who also became involved in the CYC. Among them were James Wah-Shee and George Erasmus. Puxley left the CYC in 1971 to work as a researcher with Central Mortgage and Housing Corporation, but in 1973 he joined the staff of the Indian Brotherhood at the invitation of its then president, James Wah-Shee. Two years later, Puxley was one member of the Brotherhood staff who was said to be instrumental in removing Wah-Shee from office and aiding in his replacement by George Erasmus. Puxley and Iveson were still on the Brotherhood staff until late 1977, when Erasmus abruptly fired the last of his remaining white advisors.

It was Puxley who reportedly did much to steer the philosophical course of the Brotherhood, preaching against what he later told Judge Berger were "colonial patterns ... of domination, of exploitation, of oppression," and advocating a new society which will free us all from "the corporations whose imperatives define our choices."

A more direct link between Church House and northern native peoples was forged in 1970-72 when the Anglican Church provided ten thousand dollars in grants to the Nishga tribe of northwestern British Columbia to help support their appeal to the Supreme Court of Canada against the provincial government regarding their right to aboriginal title. The historic appeal to the Supreme Court, which caused Prime Minister Trudeau and the federal cabinet to reverse their policy of not recognizing aboriginal rights, was handled by a Vancouver lawyer, Thomas R. Berger.

Yet another link was provided by Peter Russell, a professor of political science at the University of Toronto, who worked closely with the people at Church House while acting as a volunteer consultant to the Brotherhood on constitutional matters and appearing as a witness before both Judge Berger and the National Energy Board. Russell also worked with Donald Simpson of the University of Western Ontario as co-chairman of the Brotherhood's Dene Southern Support Group. Established in 1974, the Southern Support Group served as the official arm of the Brotherhood in southern Canada. Its job was to explain the Brotherhood's position, seek active support (by such means as presentations to the Berger Inquiry) from sympathetic groups, raise funds, lobby politicians and cabinet ministers, and feed back information to the Brotherhood's Yellowknife headquarters.

But the focal point of the efforts of the churches for a Mackenzie Valley pipeline moratorium eventually became Project North, described as an "Interchurch project on northern development." Established in September, 1975, by the Anglican, Roman Catholic and United Churches, Project North was later joined by the Presbyterian, Lutheran and Mennonite churches. With the active participation of headquarters staff from at least the three major churches, much of the work of Project North was carried out by two staff co-ordinators. One was Hugh McCullum, a barrel-chested man with a giant beard who had previously worked as editor of the Anglican newspaper, *The Canadian Churchman*, while the other staff co-ordinator was his third wife, Karmel. The McCullums worked with boundless energy, criss-crossing the country from sea to sea and north to south, promoting a pipeline moratorium which in turn was supposed to give birth to the Dene Nation. A prolific publicist, Hugh McCullum also served as the official news media spokesman for the Brotherhood.

From Church House, meanwhile, came a steady flow of funds averaging according to a church paper, seventy-five to one hundred thousand dollars per year, "to native organizations in support of social, economic, legal and political development projects." In June, 1975, the annual General Synod of the Anglican Church, meeting in Quebec City, passed resolutions supporting native peoples in their quests for just land claims settlements and urging the government to "halt planned development until aboriginal claims are settled."

The Canadian Catholic Conference of Bishops in their 1975 Labour Day Message, which got front page treatment in several major newspapers, accused governments and corporations of having "secretly planned and suddenly announced the construction of large industrial projects." The bishops claimed that "the Canadian North is fast becoming a centre stage in a continental struggle to gain control of new energy sources." This gave rise to fears of "colonial patterns of development, wherein a powerful few end up controlling both the people and the resources." The message was signed by six Catholic bishops, none of whom were from the north. The following March, six representatives from Canada's five major churches (again, no representatives from the north) plus the Canadian Council of Churches, met with Prime Minister Trudeau and seven cabinet ministers to once more press the demand for a moratorium on "all major resource development projects in the Northwest Territories, until land claims were settled." Their demands were outlined in a document entitled "Justice Demands Action".

By this time, Judge Berger's staff was hard at work on the detailed planning and arrangements required to take the Inquiry on its national tour across southern Canada. It proved to be a news media extravaganza. It travelled to Vancouver, Calgary, Edmonton, Regina, Winnipeg, Toronto, Montreal, Ottawa, Halifax and Charlottetown to hear what the people in the south had to say about a pipeline from the north. The position of the churches on the issue was outlined in the resolutions and other papers from the 1975 Anglican Synod in Quebec City; in the Labour Day Message of the Catholic bishops; in the document, "Justice Demands Action," presented to the federal cabinet, and in a hefty book, *This Land Is Not For Sale*, written by Hugh and Karmel McCullum and published by the Anglican Book Centre. To make sure that the message was not lost, teams from Project North, the Southern Support Group, and the head offices of the major churches fanned out across the country, preparing local church leaders and groups for their appearances before Berger.

In the four weeks of his southern tour, Justice Berger heard some four hundred submissions from environmental, public interest, political, and native organizations; from labor unions, businessmen, local politicians and individuals; but more than any other group from scores of church organizations and clergymen.

When the southern hearings opened in Vancouver on May 10, 1976 the tone was set by the first witness, Rev. Wes Maultsaid, education officer with the Inter-Church committee for World Development Education. Rev. Maultsaid presented the same arguments, couched in the same phrases, that sprang from the pages of the various church documents and papers. These same arguments echoed from church people in city after city as the Berger caravan toured across the country. "The present world order is characterised by the maldistribution of wealth and control of resources by a small minority . . . The economic system in which most Canadians prosper is the same system which creates poverty." We must "insist that colonial patterns of development not prevail in the north." We must ensure that northern resources "meet human needs and not simply the interests of the transnational corporations."

The major presentation of the churches, however, was not to come until three weeks later, on June 3, in Ottawa. Rev. Dr. Russell Hatton, national consultant at Anglican Church House, read the brief of Project North, purporting to speak on behalf of fifteen million members of the six supporting churches. Sitting at a table in the former railway station where the big federal-provincial conferences are now held, Dr. Hatton was flanked by Archbishop Scott, Monseigneur Adolphe Proulx, Roman Catholic Bishop of Hull, and by Rev. Dr. Donald MacDonald of the Presbyterian Church. Before reading the submission, Dr. Hatton introduced the staff of Project North: Dr. Tony Clarke, Canadian Catholic Conference, Ottawa; Reverend Edward Johnston, the Presbyterian Church, Toronto; Mrs. Elizabeth Loweth of Toronto, Rev. Dr. Clarke MacDonald, Toronto, and Don Shepard, Regina (all these from the United Church); Rev. Ernest Willie, Toronto, Anglican Church; Hugh and Karmel McCullum, Toronto. A total of twelve names were identified with the Project North brief but not one of them from the north.

The purpose of Project North's brief, explained Dr. Hatton, "is to call for a moratorium—a minimum of 10 years has been suggested—on all major resource development in the North, including the Mackenzie Valley pipeline." This was meant "to stop any further development, exploration, drilling or the issuance of permits of any kind, until all northern native land claims have been justly settled."

"Most of us live and benefit from a socio-economic situation which is sinful," Dr. Hatton read. "We create and sustain social and economic patterns of behaviour that bind and oppress, give privilege to the powerful and maintain systems of dependency, paternalism, racism and colonialism."

The brief denied that "our society, as it presently operates, is basically sound," and found that "simple tinkering and patchwork will not suffice ... We are talking about more than simple reformism and calling for more than mere individual conversion. We are calling for a conversion within our social and economic structures ... To bless the established order is to remain unconverted."

"We must learn from our native brothers and sisters that our land and its resources are to be used for the benefit of all people and not exploited for the profits of a few," Dr. Hatton continued. The brief found "striking parallels between the struggle of the native peoples in the Amazon region of Brazil," where "a colonial pattern of resource development has emerged" and concluded that "the same forces and the same patterns ... and the same transnational corporations ... are at work in the Mackenzie." The Brazilian experience, Project North predicted, "will occur in the Northwest Territories during the next decade if the plans, developed thus far in secret, of the federal government and the transnational corporations are allowed to proceed unchecked."

(In secret? Plans to research the feasibility of a Mackenzie Valley gas pipeline were publicly announced and widely reported as early as 1970. The results of all of those studies, comprising tens of thousands of pages, were all made public, and were the subject of the most exhaustive public inquiry of any industrial undertaking ever proposed in Canada. The pipeline applicants lived in a public goldfish bowl. Not once was any public interest group denied any information requested from Arctic Gas.)

"We cannot afford to be complacent," said Project North, "about the protection of the democratic rights of the native people of the north," faced as they are with so many pressures. Among these pressures was listed "the 'colonial' administration of the territorial council," ignoring the fact that the Legislative Council of the NWT is a fully-elected body, the majority of the voters are native people, and the majority of the elected councillors are native people. But since the 'colonial' council has so far only limited legislative powers, Project North held that the native peoples' "desire for self-determination and control of their own destiny ... can only be achieved ... through economic development that they control and administer."

The brief did not mention that the rights sought for the Dene Nation included the exclusive power to determine proposed undertakings of a national and international nature, rights of a kind which are exclusively those of the sovereign authority of Parliament, rights that exceed even those enjoyed by provincial governments.

The brief touched upon the no-growth philosophy advocated in so many church statements, stemming from concerns about an increasingly fragile ecological system, resulting from our "high energy consuming, materialistic, hedonistic lifestyle."

Finally, Project North brought its expertise to bear upon an analysis of Canada's complex energy situation. It purported to demonstrate how gas reserves in western Canada could meet Canada's needs until at least the year 2001, thereby eliminating the need to build a pipeline from the western Arctic. This would be accomplished by regulation of domestic energy consumption to meet conservation goals and by diverting to Canadian consumers seventy per cent of the gas supplies that had been authorized for sale to the United States. The energy policy advocated by Project North concluded by calling for "the gradual reduction of oil and gas exports to the United States and, instead, the export of energy at below international prices, to underdeveloped countries."

Unexplained in this proposal is how we could divert energy from American consumers to meet a portion of our own needs and still have it to sell to underdeveloped countries; how this squares with a ten year deferral to tap our Arctic energy resources; or who would pay to develop energy that costs much more than Persian Gulf oil in order to sell it at a lower price. As long as they rely on materials as far removed from reality as the Project North brief, it is hard to see how the churches can aspire to make a serious and positive contribution to debate on matters of vital national importance. Yet it was papers like this one that helped create the political climate which in the end determined the outcome of the pipeline applications.

The next stage of church involvement in northern native issues related to a disagreement between the Indian Brotherhood and the Metis Association of the NWT. The two groups did not see eye to eye on the terms of a land claims proposal, and Ottawa withheld financing of the associations until they could develop a single position. The Anglican and Roman Catholic Churches stepped in to lend the Brotherhood $100,000, without any consultation with the Metis, enabling the Brotherhood to complete and ratify their Dene Nation proposal, which was duly presented to Indian Affairs Minister Warren Allmand in Ottawa on October 25. On hand to witness this historic event were Archbishop Scott and Bishop Remi de Roo of the Catholic Church. Rick Hardy, then president of the Metis Association, says that this support of the Dene by the churches "is something I'll never forgive them for." He claims that this action helped drive a wedge between northern native peoples, and that the split will take many years to heal.

While the major churches drew public attention to the pipeline issues, it was left largely to small church groups to perform quite a different role.

The Committee for Justice and Liberty Foundation, an outgrowth of the Dutch Reformed Church, was founded in 1963. It existed in complete obscurity until 1976 when it earned its niche in history by an appeal to the Supreme Court of Canada, causing Marshall Crowe to be barred from chairing the NEB hearings. While Project North claimed to speak for fifteen million church faithful, CJL claimed to speak for only eleven hundred members. Its activities are financed by membership fees, donations, and "from time to time an offering of a gift" from individual church congregations, according to its Executive Director, Gerald Vandezande. However, Vandezande told the NEB, "none of the churches have a direct voice in the decision-making process of the CJL. None are members."

CJL operates with a small staff in squeaky-floored offices above the Institute for Christian Studies, Cosmopolitan College, in downtown Toronto, where its research and policy director, a young sandy-haired lawyer from Edmonton, John Olthuis, wears construction boots and wool shirts. The principal thrust of CJL appears to be towards a return to a simpler society with a more basic lifestyle, eschewing an energy and capital intensive economy in favour of a more labour-intensive economy.

CJL's main forum was before the National Energy Board. In an internal memo outlining the issues which he hoped to raise at the hearings, Olthuis had written in 1975 that the decision on the pipeline applications "will either reaffirm a high energy consuming, economic growth maximizing way of life, or point toward a new set of values geared to human—as opposed to economic—growth." On a stage crowded with some of the highest-paid lawyers in Canada, few performed with greater skill than the boyish-faced lawyer who spoke with an Alberta twang. And few played a more effective role than Olthuis, whose appeal to the Supreme Court on the Marshall Crowe matter was one of the pivotal points along the decision path.

It was the CJL which provided the final pipeline forum for the churches, calling sixteen witnesses to appear before the NEB in the later phases of the hearings in April, 1977. Included were the Anglican Primate Archbishop Scott; two bishops and five other representatives from the Social Affairs Department of the Canadian Catholic Conference; Gerald Vandezande, representing the CJL, and Hugh and Karmel McCullum of Project North. Yet another cluster of southern church people, and again, without one voice from the north. In addition to these church representatives, other witnesses called by the CJL included Dene consultants Mel Watkins, Peter Russell and Don Simpson; Meyer Brownstone, chairman of Oxfam Canada (an organization that directed part of the money raised by school children in the Miles-For-Millions marches in aid of the starving to assist the "political development" of the Dene); and a University of Guelph professor, Gustav Van Beers.

In the informal phase of the Berger hearings, where the church people had previously testified, there was no cross-examination. But before the National Energy Board, it was a different matter. However, the lawyers were not at all certain how to deal with this phalanx of ecclesiastics. Lawyers before the board are accustomed to grilling engineers, geologists, accountants and economists, often hoping to discredit their testimony and the cause they represent. But what do you do with a bishop or an archbishop? Not very much, the lawyers decided. Seldom have witnesses before the board been treated to such limited and gentle cross-examination.

THE CHURCH CRITICS

The call for a ten year moratorium did not come from northern churchmen, it came from the south. It came from southern people with deep feelings of genuine compassion who believed romantic myths about the north and whose sense of justice was outraged over the way our society has in fact treated native peoples. But in the background of this call for a ten year delay can be plainly heard the voice of an emerging "political theology," a voice whose tenor is significantly affected by Marxist doctrine.

In the testimony of the one thousand people who appeared before Berger in twenty-eight northern communities and ten southern cities, in presentations before the National Energy Board, or in the multitude of other church papers, documents and books, I have not been able to locate one clergyman from the NWT or Yukon who endorsed a moratorium. I've read more than twenty thousand pages of Berger and NEB transcripts, and if there were any northern clergymen calling for a delay, they have escaped my attention.

Certainly, the northern clergy have fully endorsed a just settlement of land claims. They have urged that these claims should be settled before any pipeline is built. But on the issue of a lengthy moratorium, most have remained publicly silent. I suspect this reflects not so much a reluctance to challenge head office church positions as a fear of exacerbating dangerously tense racial feelings which had built up in the Mackenzie Valley. The relatively few public expressions by northern church people that I have been able to find have all contradicted the position taken by their brethren from the south.

Two Presbyterian ministers who visited Inuvik and other Delta area communities failed to find a single northern clergyman who supported the moratorium. These men were Rev. George A. Johnston, Superintendent of Missions for the Church in Alberta, and Rev. Dr. Alex F. MacSween, Toronto. They attended the sixth annual Arctic Summer School, sponsored by the University of Alberta, at Inuvik, June 18 to July 2, 1976, and wrote separate reports on their findings to their church's Board of World Missions.

"In discussions with northerners, not once was support voiced for such a moratorium," Dr. Johnston wrote. "Instead there was amazement at the ignorance of the Church and criticism of the stand. Both Roman Catholic and Anglican clergymen in the north declared they had not been consulted."

Rev. MacSween wrote that "despite diligent enquiries of the few clergymen we met in the Arctic, we did not find any who were contacted with reference to the moratorium proposal." He wrote that when they visited Father LeMeure, a Roman Catholic veteran of thirty-six years of Arctic service, they found him "very much upset by the call . . . for a 10-year moratorium" and "disturbed by the fact that there had been no consultation." MacSween quoted Father LeMerue as stating: "Do the Canadian churches want my people to starve? . . . My people all want to see development come to the north soon, and of course they want a part of it."

Dr. Randall Ivany, Ombudsman for the Province of Alberta, a Canon in the Anglican Church, told me that he too failed to find any northern clergy who had been consulted about or who supported a moratorium during his summer of 1977 visit through Mackenzie Valley and Delta communities.

Project North's brief to the Berger Inquiry resulted in a strongly worded letter to Anglican Church House in Toronto from the president of the Woman's Guild of the Church in Inuvik. The

160

author was Ms. Suzie Huskey, a Loucheaux Indian, interpreter with the Territorial Government, and a member of the town council:

> Our Inuvik Church has not been consulted and I doubt if other Churches have been, [Ms. Huskey wrote]. What do you want us to do for ten years? Pray?... A lot of people here work and appreciate the money they earn. We are not against development but do seek a just land claim. A pipeline without a settlement would not be fair to northerners. But we want to participate in the land claims settlement just as we also want to participate in and control northern development.

Ms. Huskey aimed a blast at the NWT Indian Brotherhood for failing to communicate with native people who, she said, had not "heard what sort of land claim is being proposed for us." She also said that Indian people in her area, "don't know what the Dene Declaration is about... We resent the churches donating $140,000 to the Brotherhood when we see so little native participation." She concluded that, "a hurried land claim settlement without the understanding and agreement of all Indians concerned could hurt us badly. A ten-year moratorium keeping people on welfare would also hurt us."

The three territorial members of the United Church's Northern Co-ordinating Committee, Rev. James Ormiston of Yellowknife, Doug Billingsley of Inuvik, and Charles Pearson of Whitehorse, spoke out and issued statements and press releases giving their position. "The concept of preserving a traditional, land-based lifestyle may have idyllic appeal to southern Canadians, but it is not based on the reality facing a significant number of northerners," they stated. "Northern young people have been educated in expectation of a wage economy and many of their parents are quick to point out that many young adults have little interest or desire to follow the harsh uncertainties of living off the land." The northern clergymen urged their church to, "Set Project North free to find its own support from the agencies which openly espouse its political biases" and to "abandon all support for the divisive political influence of the Indian Brotherhood of the NWT."

Few critics of Church House have been more outspoken than Alberta's Ombudsman, Dr. Randall Ivany, former Anglican Dean of Edmonton, a native Newfoundlander who has served his ministry in both Canada and the United States, and a former electrical engineer.

In an extensively quoted address in late 1976 to an Anglican Church group headed by the Bishop of Alberta, Dr. Ivany accused the church of taking a confrontation course on most matters without listening to both sides of a controversy. "That kind of simplistic, school boy approach will do nothing to assist dialogue, or help bring opposing sides together," he told the bishop and his committee. He called on churchmen to stop condemning "academics for their arrogance, business leaders for the concern for profits, government leaders for their power-seeking, and science and technology as the source of Godlessness and social ills. Churchmen should take leadership in bringing these resources together to confront and resolve the myriad of problems we face, and they must do it in love, not anger."

Several months after the pipeline decision, I visited Dr. Ivany in Edmonton. We talked for several hours in his office and over lunch. "My biggest problem with the Church is its confrontation attitude on every subject that you want to mention," he told me. He said that such actions do not really represent the church. "It's a group of people who have established themselves in Church House. I have some questions about these people. I have some very serious questions about them, as to who they are, what they are, and their whole philosophy."

"I have defended the Primate. Basically, I like the Primate. I have found him to be a very reasonable, intelligent human being who is warm and considerate and honest. I'm not convinced that he is not being led down the garden path by these people."

Dr. Ivany says he agrees with the Primate's premise that all social actions have political connotations, "and the Church has to accept that. But when we go beyond that to confrontation politics, to the attack on government and big business, and small business, too... Every statement that comes out of Church House now has to be accepted as a given that it's a left-wing statement... Certainly some of the things we are getting, and that we've got out of General Synod, are Marxist doctrine."

Listening to this man—dark, bushy eyebrows on a broad face under a mane of grey hair—you know that he is pained and uncomfortable in his role as a critic of his church. "My wife says to me, 'for God's sake stop it, because the general public doesn't

162

realize what you're even trying to do. All they see you doing is attacking the Primate.' And that's sacrosanct. You don't argue with the Primate in public." He claims that people have said to him, "Who do you think you are? Why would you attack the Primate? Or the Church?" He says his actions are seen "in the context of attacking God. You don't do this sort of thing. Even in 1977, in Canada, you don't do this."

"Many of the Church people I know, friends of mine, have said, 'look, to hell with it. I don't have to put my money there. There are lots of things I can put my money into. I don't have to go and listen to this 10-minute social service homily on a Sunday morning which, if prepared at all, is ill-prepared. It is supposed to have relevance to the Old Testament but it doesn't tell me anything about the God I worship! And so you get the dropout." He says that dropouts from the Anglican Church are being picked up by other churches, such as the Pentecostal.

"I can tell you quite frankly that if I were to apply tomorrow for a job in the Church there isn't a hope in hell that I would get one. I don't think there's a Bishop in the country who would give me a job right now, because of my position, because I've chosen to be honest and say, 'I don't believe what you're saying. I don't believe the type of attitude that you are taking is the right one. I don't believe that you are representative of the Anglican Church.' And you mustn't say this. We mustn't argue in public. It's anathema. You can't do that kind of thing in the Church. Which gives you some idea of where the Church is, or isn't."

Dr. Ivany told me that he sometimes wakes up in the night in a sweat of uncertainty, worrying whether he might be causing some people to lose their faith because of the position he has taken, and whether he is really right. "But I don't think Ted [Scott] does. I don't think Berger does. I think that they are so convinced that they are right in this kind of approach, that there is no alternative. And that scares the hell out of me."

"I have always felt the Church must be a reconciling body, a healing community. What I have seen in the last few years from Church House in Toronto, primarily from there, is a disrupting force. It is a group of people who are hell-bent for destruction, as far as I'm concerned, who have done the Church an enormous amount of harm, who have split friends and families, who have been anything but a healing community."

163

Although he says the Primate denies it, Dr. Ivany pointed out, "I don't think the Primate is in charge. I think the Primate is being manipulated by a very bright bunch of people who know exactly where they're going and what they want to do. And they're using the Church to do it."

He feels that most of the Anglican community, who "deeply love God and their Church, don't understand anything of this that is going on. By the time they discover it, if they ever discover it, it's going to be too late... I don't think there's any turning back. I don't envy the next Primate."

Dr. Ivany said that at the Anglican 1977 General Synod in Calgary, only he, one other clergyman and a few laymen, opposed certain Church positions and statements. "It's like a guy running up the side of a mountain, and looking behind him, suddenly discovering there's not a damn soul there."

"I guess the big difference between myself and the Primate and his group is that I am not completely sure that I am right. They are. I think there is built in a good deal of margin for error in the kind of approach I have taken. I believe that I am right. I see the hurt as a result of some of the things that have happened. I see people leaving the Church in droves. I see friends and families taking opposing views. And that causes a good deal of hurt. And I think to myself, 'this is not the Church's responsibility or the Church's job'. But there is always that margin for error that just maybe... But I've never seen any bending on the other side. The Primate talks of not wanting to get ourselves into that kind of a corner where we're taking attitudes that we can't go back on. Which is precisely where they are, and have been for some time now."

In Toronto, I spoke to Dr. Russell Hatton, former national consultant at Church House and the Anglican's principal representative on Project North, who is now with the Toronto School of Theology. I asked him about the political direction of the Church, and what it meant by its new economic order.

"Often socialism and Christianity get equated in people's minds when you begin to look at what you call a new social order," Dr. Hatton explained. "The churches are not trying to press for an ideology, a particular type of social order. I don't think you're going to find in the churches any agreement, you certainly won't among the staff, any consensus about that. The church, from its political theological position of doing a critique of society, is pointing to certain need for change to take place. It is looking for a new order which comes from a Biblical theological position that our Lord was calling for a new creation. There was never any discussion in my circle about what that new order would look like. But there are some values that underlie how to work out such an order. In every society, whether it is capitalist or socialist or communist, the church is called to critique that order."

Dr. Hatton says that the churches are calling for greater participation by people in the decisions affecting their lives, as for example by the Dene in the Mackenzie Valley, who he says are struggling for self-determination, to find their own identity and retain their culture. He also says that they don't see that being accomplished "apart from the political decisions that are being made."

He claims that if enough people participate in shaping society, "then perhaps we can find a new order that provides a more equal basis. That's what's going on. If people want to translate that into an ideological position, then I suppose that's their perception."

"Some of the new patterns that have emerged aren't any better than the old patterns... I've been in Roumania and I wouldn't want to live in Roumania, or live with that kind of government... The colleagues I work with, we deplore the communist ideology just as much as we deplore the dictatorships of the right wing ideology ... If the NDP party went into power tomorrow," the churches should be exactly in the same position in respect to critiquing that government as any other. "I don't care what party is in power."

"What the Church was doing in the Mackenzie Valley was that it had made up its mind that the native people had a right to be heard in the fullest possible way, and that there was an imbalance in the forces and power patterns. The natives needed to be stood beside, so that what ever happened, they would have an equal opportunity at that."

Dr. Hatton believes that the decision not to build a pipeline along the Mackenzie Valley has provided the native people with what they sought: time to work out their own self-determination. However, one wonders whether, now that the churches have helped provide the time, they will also assist in the more difficult task of establishing self-determination for those who are out of work, the day to day situation of most native people in this region?

"The church's work in standing beside the native peoples in the Mackenzie Valley has just begun," according to Dr. Hatton. He noted that the council of the North had already started to look at this challenge.

"If the church has any good news at all, it's the good news that we can live together, we can work together, we can build up a new society based on values of participation and equality and somehow helping the natives who, in the end, have to make their own decisions. They have to develop themselves. If they don't take this opportunity they have got only themselves to blame."

THE PRIMATE

It was on April 1, 1974, that I first met the Primate of the Anglican Church of Canada. Four of us from Arctic Gas, led by Bill Wilder, had met at Church House for two and a half hours with the church's Unit on Social and Public Responsibility to answer whatever questions they might have about the proposed pipeline. The unit had been previously briefed by Glenn Bell, counsel for the Indian Brotherhood, and was at that time (this was before the formation of Project North) planning to make a submission to the Berger Inquiry.

The Primate, who had other commitments, was able to attend only a part of the session. There were thirteen other church representatives, among them Dr. Hatton of Church House; Tony Clarke of the Canadian Catholic Conference; NDP Member of Parliament Andrew Brewin; Dr. Peter Russell of the University of Toronto, co-chairman of the Indian Brotherhood's Dene Southern Support Group; and one native person, William Kay. No one from the north was present, except one of the Arctic Gas representatives, Tim Taylor.

According to my notes of the meeting, this group appeared to receive us with considerable mistrust of our motivations and suspicion of our arguments, while supporting the no-growth philosophy that they felt was being imperiled by the pipeline. My notes also record, however, that "the single factor with which this group was by far most concerned—including the Primate—was the settlement of native land claims . . . Indeed, it was explicitly suggested that the pipeline is the lever to achieve a long over-due settlement of this issue, and that if necessary the pipeline should be delayed in order to force the attainment of this settlement."

My notes also refer to the brief statement made by the native representative at the meeting, Mr. Kay. "He asked if we recognized whether native people 'have any rights'. And he warned of possible violent action by native peoples if the pipeline were to proceed across their land and against their will. He said northern native peoples have a supply of more than one thousand rifles assembled for possible use in militant action; that they know the effect of, and how to employ guerrilla tactics. He alluded to the possibility of violent action as early as this summer [1974], but did not suggest what might trigger this."

But my most lasting impression of this session was of the closing words of the Primate, from which we at Arctic Gas took encouragement. At considerable length, he argued against the adoption of rigid and dogmatic positions which could later make it difficult to find any accommodation of differing views and interests. It seemed to be a pretty direct message to those on the committee who were involved in planning the submission to the Berger Inquiry and who were later involved in the Project North effort.

Two years later, when Project North finally presented its brief to Justice Berger, I was left with the impression that no one had paid the slightest attention to what the Primate had said that day.

It was several months after the pipeline decision before I met the Primate again, this time for an interview in his unpretentious office on the third floor of Church House. A surprisingly diminutive man, Ted Scott struck me more like everyone's favourite uncle than "the most influential figure in Canadian religion." He was dressed not in scarlet robes, nor even clerical garb, but a plain dark suit and tie and a grey wool sweater. While we talked, we drank coffee from paper cups.

I asked the Primate why the call for a moratorium had not, so far as I had been able to ascertain, been publicly advocated by any of the clergy from the north.

The pipeline moratorium, he said, had to be viewed in the context of an advocated moratorium on all northern development, pending settlement of native land claims. This call had been issued, prior to the emergence of the pipeline issue, as a result of representations by northern clergy and native people concerned with the hydro development project at James Bay, and also in reference to the Nishga Indian people in northern British Columbia who had been seeking a settlement of their land claims for more than a century. Who stated the church's position was less important than how it had been arrived at. In any event, many of the presentations before Berger endorsing the pipeline moratorium had come from church people in the community who had identified themselves with their communities rather than as church spokesmen.

The settlement of native land claims, in Archibshop Scott's view, "is one of the crucial questions which we believe the whole future of Canada relates to." He says that northern development is taking place without the agreements with native peoples which are called for under the British North America Act. "We have not honoured the legal obligations that are a part of the basic heritage of this country." He feels that corporations also bear some responsibility to see the claims settled, a responsibility which he does not feel corporations have met. "On the one hand you are saying that you want that [settlement of land claims] done, and on the other hand you are applying for all kinds of rights without that being done."

By thwarting new employment opportunities, won't the pipeline decision contribute to social disintegration in the Mackenzie region? Archibshop Scott suggests that it is difficult to measure the type of social disintegration that takes place under either of two patterns—lack of employment opportunities, or the "great deal of social disintegration that has always taken place when you put development projects alongside native communities."

What is meant by the new economic order? "I don't think anybody can document that in detail at the present time," the Primate says. He sees it as an attempt to bring new values to bear on what he perceives as a range of problems facing the whole world: an increasing gap between the poor and the prosperous, steadily increasing use of non-renewable resources resulting in the risk of an unsustainable society, the need for greater participation by people in the decisions that shape their lives.

"The division that we tend to make between socialism and capitalism is no longer a valid division. . . . I would opt for the kind of situation that provides the maximum freedom to individuals." The new economic order "is a way of responding to a new kind of challenge that has become clearer in the last forty or fifty years," a challenge that is "quite different than anything we have known in history. The traditional response of socialism, as we have known it, or capitalism, as we have known it, neither of them are adequate. The term 'new economic order' is an attempt to get out of the commitment to either socialism or capitalism, and try to get some

168

intermediate situation that would see some of the basic values of both . . . I think it is struggling for something between the two."

I asked the Primate if, as Dr. Ivany has suggested, it is really the staff members who are responsible for many of the positions and statements that come out of Church House. I noted my impression that the Project North brief had paid not the slightest attention to the advice he had given that day when Arctic Gas had met with him and others at Church House. He admitted that it might appear like this, but such is not really the case, that in fact he agreed with the basic thrust of the Project North brief. He added, however: "I was not happy with some of the language in that brief. There are certain things in it that have not happened since, and won't happen again."

CHRIST AND SOCIALISM

An empathy for socialism and an antipathy to a profit-motivated free market economy is not a new element in Canadian churches. A 1918 report of the Methodist Church declared that the First World War "has made clearly manifest the moral perils inherent in the system of production for profits" and that, "methods based on individualism and competition have gone down like mud walls in a flood . . . We do not believe this separation of labour and capital can be permanent. Its transcendence, whether through co-operation or public ownership, seems to be the only constructive and radical reform."

The first socialist elected to the House of Commons, in 1921, was a former Methodist minister, James Shaver Woodsworth. He was later elected the first leader of the Co-operative Commonwealth Federation, whose founding Regina Manifesto (1934) declared: "No CCF government will rest content until it has eradicated capitalism and put into operation the full programme of socialized planning which will lead to the establishment in Canada of the Co-operative Commonwealth." (The New Democratic Party, successor to the old CCF, has largely abandoned this goal, and is now regarded as a party of social democrats.)

In 1933, the Toronto Conference of the United Church passed this ringing declaration: "We believe that the social realization of the Kingdom of God is not compatible with the continuance of the capitalistic system, and we think the Church should uncover fear-

169

lessly the anti-social and unchristian basis of that system and declare unremitting war on it."

It must be said that the Christian socialists and social democrats are, as individuals, among the most appealing men in Canadian history. From James Woodsworth to Thomas Berger and Archbishop Scott, they are stamped as men of deep compassion and unremitting dedication. The valid question is whether the policies they advocate will promote the goals they seek.

Some further insight into the political leanings of at least some present-day theologians was provided by Dr. Hatton, when he gave me a copy of a paper which he suggested I might find useful. It was the keynote address to a two-day seminar of some two hundred theologians on "Political Theology in the Canadian Context," held at Saskatoon in March, 1977. It was a lengthy lecture, delivered by Dr. William Hordern, president of the Lutheran Theological Seminary at the University of Saskatoon. Dr. Hordern wrote of "bourgeois capitalism," the "proletarian people," and "giant international firms ... [which] manipulate the food supply of the world in a way that leads to starvation." More significant is Dr. Hordern's account of the evolution of political and liberation theology, which he concludes are essentially the same. He reports that in the Protestant churches the movement began in Europe at the start of the 20th century with a Swiss pastor, Karl Barth, who is said to represent "the watershed for 20th century protestant theology." According to Dr. Hordern, Barth "changed the whole context in which modern theology has operated" and "overthrew the liberal theology that had dominated Protestant thought in the 19th century." In North America, at about the same time, the "social gospel" launched its attack on capitalism, marking another "break from liberalism." Later on, the liberation theology "developed in the Third World countries ... written by people who see the social struggle from the viewpoint of the exploited." As Dr. Hordern sees it, there is "no sharp line of demarcation between liberation theology and political theology."

The trouble with old-fashioned liberal theology is that it was "too much concerned with the private inner life of persons"—saving sinners, instead of saving society. The idea that the world would be a better place if each one of us adopted Christian ethics is dismissed as "pious moralisms" which are "irrelevant to the issues at stake." They "overlook the machinations such as the international agro-business corporations that manipulate the distribution of food for profit." It is not enough to "simply repent of our individual sins and change our individual ways." We must also "see the sin that is invoked when we participate in our society." With political theology we will come to see "our own contribution to the sinful state." Such as buying exports from poor countries "at a price which condemns the people . . . to a life of poverty and degradation."

Another problem with liberal theology is that it sought "universal truths that apply equally to all people," while "divine truth, like all other truth, has to be understood from particular points of view," such as the viewpoint of the class struggle. While liberal theologians "thought they were seeking the truth, the whole truth and nothing but the truth," liberation theology makes it clear that they were only "speaking the truth from the perspective of a white, male middle class."

"Once political theology has revealed the class-interest of all theology," then the rhetorical question is, from which class should it be viewed? And the answer is that "theology must be written from the perspective of the poor, the oppressed, the wretched of the earth."

Was Jesus Christ a socialist?

The liberation theologians would appear to believe so. Like they, "Jesus identified himself with the poor and promised them liberation from oppression and a coming Kingdom of Justice." He, like they, "interpreted his mission and purpose in terms of liberation for the oppressed groups in society." Thus "both the goals of Christianity and socialism are concerned with the poor and the oppressed and aim to bring social justice for them." While the actions of socialists bear watching (even a socialist may err and sin), at least "what Christianity and socialism want are very close."

And Saint Marx, too, has provided a guiding light. Dr. Hordern notes that Marx was less concerned with interpreting the world than "he was concerned to change the world," just like Jesus Christ and the liberation theologians. "A considerable number of political theologians have turned to Karl Marx for methodology to analyse society, finding a number of features in Marx that are parallel to the Bible," Dr. Hordern states.

"If a Marxian analysis of society is used and the aim is to be active in the political sphere, why not call it a new form of politics?" Dr. Hordern asks. A good question.

But political theology adapts to the circumstances, and in the Canadian context it is thought unlikely that it "could draw as heavily on Marxism as the political theology of other lands. Canada has a strong socialist tradition but it has owed very little to Marx."

So political theology has one message in one country, another message in another country. What ever happened to the everlasting, universal word? Well, there are no universal truths, says political theology, truth is what you see from your own perspective.

In contemplating the progression of conservative, liberal, political and liberation theology, it may be helpful to recall the advice of the Scottish sage, Thomas Carlyle. He taught that "a deep sense of religion was compatible with an entire absence of theology."

Perhaps the most effective response to the role played by the churches in the pipeline debate comes from a report produced in 1977 within the Anglican Church. "For the Elimination of Poverty and Social Injustice" was the result of four years of effort by the Church's National Task Force on the Economy. Drawn from across the country, the eleven members of the Task Force seem representative of the mainstream of the Anglican community. They included a member from the Church House staff, two parish priests, a businessman, a housewife, a statistician, a civil servant, a community worker, and three economists.

The report rejected "revolutionary changes in the ownership and control of industry," as well as "strict socialism [as a goal] which our Church may support." The members favoured "the widest possible ownership of private property, and so come much nearer to the idea of a property-owning democracy widely supported in post-war years throughout the English speaking world."

As for church advocacy of a new economic or social order, the report quotes Archbishop William Temple: "There is no such thing as a Christian social idea, to which we should conform our society as closely as possible. We may notice, incidentally, about any such ideas from Plato's *Republic* onward, that no one really wants to live in the ideal state as depicted by anyone else."

On the no-growth philosophy: "World poverty can only be alleviated by a vast increase in world production and a reduction, or at least a stabilization, of world population."

On calls for specific economic measures, as in the Project North brief, the report concluded that such analyses must be based on "detailed consideration of technical complexities in which individual

172

Christians might legitimately disagree, and upon which the Church as a whole is incompetent to pronounce. It is the business of the Church to lay down canons of orthodoxy in matters of faith and order. It is not the business of the Church acting in its synods to lay down detailed plans for economic and social reform, though individuals driven by Christian principles will certainly do so."

Chapter 11:
Judge Berger and the Vision
of a Northern Marxist Nation

*Let me tell you, Mr. Berger, and let me tell your nation, that this is
Dene land, and we the Dene people intend to decide what happens on
our land.*

*We have our own land, our own languages, our own political and
economic system. We have our own culture and traditions and history,
distinct from those of your nation.*

*You have forced us into communities and tried to make us forget
how to live off the land, so you could go ahead and take the resources
where we trap and hunt and fish. You encourage us to drink liquor
until we are half crazy and fight among ourselves... The Government
knows very well that liquor helps keep my people asleep, helps keep
them from realizing what is happening and why...*

*Obviously Mr. Blair and his friend Mr. Horte want to see us
destroyed. Maybe, Mr. Blair, that is because you do not know or
understand us.*

*Mr. Blair, there is a life and death struggle going on between us,
between you and I. Somehow in your carpeted boardrooms, in your
panelled office, you are plotting to take away from me the very centre
of my existence. You are stealing my soul. Deep in the glass and
concrete of your world you are stealing my soul, my spirit. By
scheming to torture my land, you are torturing me. By plotting to
invade my land you are invading me. If you ever dig a trench through
my land, you are cutting through me. You are like the Pentagon, Mr.
Blair, planning the slaughter of innocent Vietnamese. Don't tell me
you are not responsible for the destruction of my nation. You are
directly responsible. You are the twentieth century General Custer.
You have come to destroy the Dene Nation. You are coming with
your troops to slaughter us and steal our land that is rightfully ours.*

You are coming to destroy a people that have a history of thirty thousand years. Why? For twenty years of gas? Are you really that insane?

Our Dene nation is like this great river . . . we take our strength and our wisdom and our ways from the flow and direction that has been established for us by ancestors we never knew, ancestors of a thousand years ago. Their wisdom flows through us to our children and our grandchildren to generations we will never know. We will live out our lives as we must and we will die in peace because we will know that our people and this river will flow on after us.

We know that our grandchildren will speak a language that is their heritage, that has been passed on from before time. We know they will share their wealth and not hoard it, or keep it to themselves. We know they will look after their old people and respect them for their wisdom. We know they will look after this land and protect it and that five hundred years from now someone with skin my color and mocassins on his feet will climb up to the Ramparts and rest and look over the river and feel that he too has a place in the universe, and he will thank the same spirits that I thank, that his ancestors have looked after his land well and he will be proud to be a Dene.

Excerpts from testimony of Chief
Frank T'Seleie, to The Mackenzie
Valley Pipeline Inquiry, Fort
Good Hope, August 5, 1975.

The native people of the Northwest Territories chose the Mackenzie Valley Pipeline Inquiry, Judge Berger wrote in his report to the Minister of Indian Affairs and Northern Development, "to be a forum for the presentation of their case before the people of southern Canada."

Judge Berger did not let them down.

He had already done more than perhaps any other single person to advance the case for native peoples in Canada. He had led an eight-year struggle through the courts to the Supreme Court of Canada, causing the government to reverse its position and recognize the aboriginal title of native people. Now, with the Mackenzie Valley Pipeline Inquiry, he was to spend another four years and $5 million to carry the case further. He arranged federal funding for four native organizations to assist them in developing the case they were to present through his forum. He provided them with a national platform more powerful, with news media coverage more extensive, than any public inquiry in Canada had ever attained. He devoted most of the two volumes of his report to what is possibly the most eloquent plea for native claims that has ever been published, echoing virtually every argument that was placed before him by the Indian Brotherhood of the Northwest Territories (which has since changed its name to the Dene Nation). He cast aspersions on testimony which tended to contradict the case presented by the Dene, and chastised those who questioned the Dene testimony.

"At the heart of my recommendations is the need to settle native claims," Judge Berger wrote.

"Berger has written a report on native claims," Senator Jack Austin of Vancouver observed. "The pipeline is just an issue which causes the report to be written on native claims."

No one with the slightest sense of social justice would question Berger's objectives: a recapture of dignity and self esteem for native peoples and the eradication of cancerous social pathologies which are nothing less than a national shame. But if no one questions the objectives, many have questioned a great number of the means advocated to attain them.

"As a component in the decision-making process, the report is an embarrassment," in the opinion of Professor J. C. Stabler of the Department of Economics and Political Science at the University of Saskatchewan.

For taxpayers in southern Canada, one thing is certain. The ten year delay in development of the oil and gas resources of the Mackenzie Delta, urged by Judge Berger, will be costly. During that decade southern taxpayers will have to subsidize the government and economy of the Northwest Territories by probably more than five billion dollars, while billions of dollars in potential revenues from Delta gas are deferred.

Judge Berger has also provided, if unwittingly, a powerful advance for those who seek to establish a Dene Nation as a separate Marxist state in the north.

"There is no question whatsoever that his interpretation of events in the Northwest Territories does lend confidence and credibility to some of the most radical elements in the Northwest Territories," Senator Austin claimed in the *Vancouver Sun*, June 3, 1977.

In addition to Judge Berger's unwitting assistance, the Dene Nation leaders had also received five million dollars in federal

176

government funding by 1978, money which was used to advance the cause of a Marxist state, although it certainly was not granted for that purpose.

THE JUDGE

Thomas Rodney Berger, son of an RCMP sergeant, grandson of a Swedish judge, has already changed the entire approach of the Government of Canada on aboriginal rights. Now he wants to change it more, much more.

Born in Victoria in 1933, Berger attended law school at the University of British Columbia. He is remembered as a sombre, hard working student who spent his summers working in the sawmills and logging camps of British Columbia, graduating third in a class of sixty. He hung up his legal shingle in Vancouver, where his mother worked as his bookkeeper, and says he learned much about law in a beer parlour from his mentor Tom Hurley, a Vancouver criminal lawyer with an eloquent Irish tongue.

"He had a great magnanimity," Berger recalled of Hurley during a television interview in 1977. "He had no money. He drank an awful lot. He was the dean of the criminal bar in Vancouver... I got into the habit of dropping in to see him to talk about the cases I had coming up. We would adjourn to the beer parlor. I learned a lot of law and trial work in the beer parlor with Tom Hurley. And he never grudged me the time... He was a great influence on me ... and his generosity to anyone who came to his office who needed help in the courts... He was never concerned about money, never. In fact, he died without a penny. That was, in a way, adhering to the finest traditions of our profession."

Another major influence, writes Allan Fotheringham in *Weekend Magazine*, was Maisie Hurley, a cigar-smoking firebrand who married Hurley after he had been a bachelor for sixty-four years. Maisie prodded both her husband and Berger to take up the cudgels for B.C. Indians. She was, writes Fotheringham, "this striking, strident woman raging about this cluttered Vancouver apartment," a woman in her late seventies, "founder of the first Indian paper in Canada, a tireless trumpeter of her case that British Columbia was the only place in North America where a treaty was never signed with the Indians," (not quite—there was also Alaska, much of Quebec, and nearly all of northern Canada).

Berger worked for clients whose cases pitted him against the establishment, and often against government, clients such as Indian bands, the International Woodworkers of America, miners suffering from silicosis, a civil servant who brought a slander action against B.C. Premier W.A.C. Bennett which Berger took to the Supreme Court of Canada and won. He once said of the establishment, according to *Maclean's*: "I've taken on the bastards and beaten them."

His reputation as a legal establishment beater led to a relatively brief career in politics, running unsuccessfully as an NDP candidate in the 1960 provincial election and successfully in the 1962 federal election. His political opponent, Ron Basford, later federal justice minister, called him "a man dangerous to the people of BC." In 1969, Berger won the leadership of the provincial NDP party, but in the provincial election four months later, lost his seat while the NDP party suffered a shattering defeat. He decided it was time to quit politics.

"I think getting thumped in an election in full view of everyone in the province does something to increase your supply of humility," Berger recalled in the 1977 TV interview. "I think it prepares you for whatever shocks may come later on in life."

Dave Ablett of *The Vancouver Sun*, who covered the 1969 election, told me that Berger was known in the press corps as "the only politician who always looks at every question from all six sides."

Berger likes to ponder matters more deeply than the pressure of politics allows. "I was always uncomfortable as a politician," he told Mike McManus during the TV interview. "I found, heading a royal commission...you have a chance to examine a problem at length, to bring the best thinking you are capable of to bear on it, to make a judgment, to make recommendations. Intellectually that's a good deal more satisfying than politics, because in politics you're often rushing after instant opinions on so many things. I was always unhappy about that. I knew it had to be done, but it's not terribly satisfying intellectually. The other thing I found was that I wasn't very good at meeting people."

Berger's most important case began in 1966 when he launched an action against the Attorney General of British Columbia on behalf of the Nishga Tribal Council, seeking a declaration that the aboriginal title to ancient tribal territory covering more than a thousand square miles in northern BC had never been lawfully extinguished.

The Nishga had been seeking legal title to these lands for more than fifty years. In 1913, they launched a petition to the Privy Council, followed by further petitions to special commissions of the Senate and House of Commons in 1922 and 1961. Berger based his case on the proclamation issued in 1763 by King George III which, in essence, disallowed the possession of Indian lands without prior settlement between the Crown and the Indian people involved.

"Lawyer Says Indians Own BC," screamed the headline in *The Vancouver Sun*. Most of the legal fraternity dismissed the idea out of hand. Justice Gould of the British Columbia Supreme Court dismissed the action in 1969, ruling that "any rights the Indian bands had in the lands in question were totally extinguished by overt acts of the Crown Imperial." The verdict was upheld by the BC Court of Appeal the following year.

Not only the courts, but the federal government as well had rejected the concept of aboriginal title. A federal statement on Indian policy in 1969 declared that "aboriginal claims to the land ... are so general and undefined that it is not realistic to think of them as specific claims capable of remedy except through a policy and program that will end injustice to Indians as members of the Canadian community." Asked about aboriginal claims at a meeting at Vancouver in August that year, Prime Minister Trudeau responded: "Our answer is no. We can't recognize aboriginal claims because no society can be built on historical 'might have beens'."

Undaunted, Berger took the matter to the Supreme Court of Canada where, in January, 1973, he obtained the ruling that was to completely reverse Trudeau's policy. Three of the Supreme Court Justices, Laskin, Hall and Spence, held that the Nishga did, in fact, have an unextinguished aboriginal interest in the land. Three other justices, however, held that the Crown had acquired "absolute sovereignty over all the lands of British Columbia, a sovereignty inconsistent with any conflicting interest, including one as to 'aboriginal title'." The appeal lost on a technical point considered by the seventh justice, rather than on the issue at hand.

Following the Supreme Court decision, Trudeau allowed as how perhaps there was such a thing as aboriginal rights after all, and that they would have to be dealt with. Seven months later, Indian Affairs Minister Jean Chrétien spelled out the government's reversed position. This major policy statement dealt with the claims of native people which "relate to the loss of traditional use and occupancy of lands in certain parts of Canada where Indian title was never extinguished by treaty or superseded by law." Chrétien stated: "The Government is now ready to negotiate with authorized representatives of those native peoples on the basis that where their traditional interests in the lands concerned can be established, an agreed form of compensation or benefit will be provided to native people in return for their interest."

In fact, the government had already been providing funds to native groups to help them conduct research in support of their claims. With this policy statement, the government recognized that native people still retained aboriginal title to nearly half of Canada, nearly two million square miles covering northern Quebec, the NWT, the Yukon and British Columbia. This interest would have to be dealt with.

While he was still preparing the Nishga case for the Supreme Court, Berger was appointed to the BC Supreme Court by then Justice Minister John Turner. The federal government, meanwhile, was seeking to settle claims on the lands which native peoples said belonged not to Canada but to them.

The first comprehensive, modern-day settlement of native land claims in Canada to follow the Supreme Court decision, was the James Bay Agreement. It involved the interests of some sixty-five hundred Cree and forty-two hundred Inuit to three hundred and seventy-nine thousand square miles in northern Quebec, sixty per cent of the total land area of the province. Agreement in principle was reached in 1974, and the James Bay Agreement became law late in 1977.

This settlement extinguished the aboriginal title of the Cree and Inuit. In return they are to receive $225 million during twenty years, exclusive hunting, trapping and fishing rights throughout sixty thousand square miles, plus exclusive rights to certain wildlife species in the remaining two hundred and nineteen thousand miles, five thousand three hundred and fifty square miles in fee simple, guaranteed minimum annual incomes for hunters and trappers, a wide range of local governmental authorities, including separate Cree and Inuit school boards.

The James Bay Agreement has been widely acclaimed as the most generous settlement achieved by native people in Canada, and strongly condemned for having been forced on the native people under the pressure of the James Bay hydroelectric project, and for extinguishing their claims to ownership of most of the land.

"I argue that the Governments of Canada and Quebec did not put the Crees to the wall to sign an agreement," Chief Billy Diamond, Grand Chief of the Quebec Council of Cree, declared before the Berger Inquiry at Montreal in 1976. "The whole objective of the Agreement is to save a culture and a society ... it gives a choice for our people in which society they wish to participate. If our people wish to participate in the traditional society, they can do so and they can benefit from this Agreement. If our people wish to participate in the modern industrial society, they can choose that path ... if our people wish to take part in both societies, it is also made available ... The Agreement puts the Cree people in a position where they will become masters of their own destiny, where they will become independent, a lot more independent than any other native group in Canada."

Regardless of the controversy, because it emanates from the Nishga case, the James Bay Agreement is indirectly a victory for Tom Berger. Beyond dispute, it is the most generous settlement that native people have won in Canada. But one victory is not the end of the war for Berger. He is one of the critics of the James Bay Agreement. "The native people of the north reject the James Bay Agreement model as inadequate to protect their traditional economy because it does not entrench hunting, fishing and trapping rights through ownership of the land," he wrote in his Inquiry report. This agreement, he says, shows "how easy it is for the dominant society to discount native aspirations whenever they are inconveniently opposed to cultural, political or industrial imperatives of the dominant society."

Having forced the government to recognize aboriginal title, Berger now asks the government to reverse another policy. Instead of settlements that extinguish aboriginal title in return for negotiated compensation, he seeks, in his Inquiry report, settlements which will confirm native ownership of the lands with legal title under Canadian law. Not, presumably, full title to all the disputed lands – what would native people do with the City of Vancouver? – but at least to vast tracts, on which they could sustain native societies and economies based on renewable resources and revenues from non-renewable mineral resources.

Berger has still another objective: the right of native people to "self-determination." This means, as outlined in the Inquiry report, that native peoples would govern themselves in all-native societies with their own forms of political and governmental institutions.

Can a judge crusade for a cause?

"I don't care who you are, you have a certain outlook on life," Berger said in a 1977 TV interview. "As a judge you are bound by the law, and you work within the law." He said that in his pipeline Inquiry he was bound by the mandate from the government, "but you look at it through your own eyes and your own experience."

To a substantial degree, the rulings of great jurists, men of profound social conscience, working within the interpretation of the law, perhaps viewing it through their own eyes and their own experience, have advanced minority rights in Canada and the United States. Berger has devoted twelve years as champion, as he sees it, of the minority rights of Canada's native people. It is difficult to imagine that he sees the task as completed.

Writing in *Weekend Magazine* on the eve of the Inquiry hearings in 1975, Allan Fotheringham asked: "Is Tom Berger ... saddled with the single most important task ever given an individual judge in this nation's young history?"

There seems little doubt that Berger has the breadth of vision (although his critics would say it is blurred), and the sense of mission, to accept such a challenge. On the final day of the public hearings of his pipeline Inquiry, the last words of Judge Berger were a warning: "You will be hearing from me." The warning still stands.

THE DENE NATION

The fire of radical activism and revolution which raged through U.S. universities in the sixties and early seventies, sparking bush fires in Canada, was all but dead by 1978. One ember, at least, still glowed in the valley of the Mackenzie River, fanned by the leaders of the Dene Nation, formerly the Indian Brotherhood of the Northwest Territories.

In the Yellowknife office of George Erasmus, youthful president of the Dene Nation, giant posters of Che Guevara and other liberation leaders, still adorned the walls. A former Brotherhood employee, writing in Yellowknife's *News of the North* late in 1975, said that in the Brotherhood's offices he had "found posters of Che Guevara, Fidel Castro and Mao's little red book being pushed onto field workers in order to propagate that faith."

A memo written in April, 1975, by Erasmus, then director of community development, to the Brotherhood's field workers, states: "Did you hear that Cambodia has fallen to the people's liberation army today? Three cheers to the movement." The memo offers the field workers a lengthy list of resource materials—books, pamphlets, posters, films—with titles such as: *The New Socialist Revolution*; *Revolution For the Hell of It*; *Mau Mau from Within*; *Notes on a Strategy for North American Revolutionaries*; *Imperialism and the Working Class*; *Tanzania Party Guidelines*.

It had taken several years for the Marxist revolutionaries to gain control of the Brotherhood. The Brotherhood had started on the pattern of other Indian organizations in Canada, as a means to redress the grievances of Indian people in the Northwest Territories.

In the late 1960s, the only forms of Indian organization in the Mackenzie Valley were the band councils in the smaller communities, and the Chiefs' Advisory Council, comprising the chiefs from these communities. The Chiefs' Advisory Council met but once a year to advise the Department of Indian Affairs and Northern Development.

The Company of Young Canadians arrived on the scene to help the Indians form a more effective organization. Among them was Peter H. L. Puxley. A Rhodes scholar, Peter Puxley had studied economics at Dalhousie University, with post-graduate work at Oxford and Clark University, and held a Master of Science Degree in planning from the University of Toronto. For three years, 1969-71, he was staff person in charge of the CYC's Great Slave Project, based in Yellowknife.

One of the first native staff members hired by the CYC, in 1968, was a young Indian who stood out as a promising leader. James Wah-shee was then in his early twenties, an Indian from the Dogrib community of Fort Rea, seventy miles west of Yellowknife. Wah-shee had been educated in the mission school at Fort Resolution and at Yellowknife where he completed high school. A promising Indian painter, Wah-shee obtained a job as promotional artist with an all-native band, the Chief Tones, which set out on a North American tour. This led Wah-shee to the University of Wisconsin where he studied business administration for two years before returning to Fort Rea and the staff job with CYC. His job with the CYC was to tour the communities in the Great Slave Lake area to stimulate interest in the development of a native organization.

The Chiefs' Advisory Council was dissolved in 1969, at the request of the chiefs, and the Indian Brotherhood of the Northwest Territories was organized with federal government funding. Wahshee was hired as vice-president with an annual salary of $18,000.

When he had returned to Fort Rea, Wah-shee found Caroline Pickles, a young blond dynamo who had been hired by the NWT government as community development officer for the settlement. The daughter of a long-time IBM employee in Toronto, Caroline, too was a graduate of the CYC. They were married in 1969. Caroline became the first secretary of the Brotherhood, ploughing into her work with tireless energy. In 1970, Wah-shee was elected Brotherhood president, following a campaign organized by Caroline that saw the two of them travel to every community in the Mackenzie Valley and Great Slave Lake region.

With the aid of government funding of more than half a million dollars a year, Wah-shee assembled a staff and a group of advisors and consultants at the Brotherhood's Yellowknife offices. George Erasmus, then 25, was hired as director of community development. After completing high school in Yellowknife, Erasmus had worked as a telephone linesman before joining the Brotherhood. Mel Watkins, the University of Toronto professor of economics and former leader of the NDP Waffle faction (later dissolved by the party because it was too socialist even for the NDP), was hired in 1974 to serve for two years as economic advisor. Other white advisors hired by the Brotherhood included Puxley, Stephen Iveson (another CYC worker who had arrived in Yellowknife from North Bay, Ontario, in 1968 to help organize the Indians), Wilf Bean (a former territorial government employee who composed ballads to relay the story of the Dene struggle), and lawyers Gerald Sutton and Glenn Bell.

Formation of the Brotherhood was soon followed by the Metis Association of the Northwest Territories, led, in 1974, by Rick Hardy, a young Metis from Norman Wells.

The two associations worked together to develop a common proposal for the settlement of land claims, united on one compelling message: no pipeline before settlement. The joint effort lasted two years, and was marked by bitter differences, before the Metis and

184

Brotherhood finally split over divergent views on the pipeline and the form of a land claims settlement.

Both associations still agreed on one point: the land claims must be settled before a pipeline is built. But after that, the Metis wanted to see the pipeline built for the jobs and economic opportunities it would provide; the Brotherhood did not want to see a pipeline built for ten to twenty years, if ever. There were differences over the increasing Marxist philosophy of the Brotherhood. Finally, there was the fact that a separate Metis organization, with different ideas, approaches and interests, stood in the path of the Brotherhood's evolving concept of a unitary native society in the north.

In his three year term as Metis president, Hardy faced tough battles fighting off proposals to disband the Metis Association and merge it into the Brotherhood. It was a fight which became bitter and venomous. While the Metis rejected a merger, the Dene Nation leaders had not given up. With the organization renamed in 1978, Erasmus was still seeking Metis members in the Dene Nation and a base to speak as a unified voice for native people in the region.

Wah-shee, meanwhile, faced a threat to his leadership from the Brotherhood's staff and advisors, a threat which he was slow to recognize. Less militant in his views than others at the Brotherhood, Wah-shee was seen as being co-opted by white interests. He had even been elected, in March, 1975, to the NWT Legislative Assembly, an alien institution in the eyes of some Brotherhood staff members and advisors.

The threat that confronted Wah-shee, and an indication of the influence of the white advisors on the emerging philosophy of the Brotherhood, was spelled out in an internal working paper prepared in 1975 by Puxley for the Brotherhood and Metis boards, which were then still working together closely, if with difficulty. The Indian movement in the Northwest Territories, Puxley wrote, "exists almost in spite of the organizations," because they had, he claimed, lost touch with the movement. "The more we adopt bureaucratic methods, the more we structure ourselves according to the government's view of the world, the more we respond to initiatives of the government and outside interests, the less we will be a part of the movement."

Puxley defined the movement as "the growing awareness of Indian people of their situation in relation to history and to the wider society which oppresses them". This movement, he wrote, is hindered by ignorance ("often fostered by the kind of education all Canadians are subject to"); by efforts to "buy off native people" ("funding of native organizations... training programs, welfare, and government employment of native people"); by "the co-option of its leadership" (while Indians face "continuing and unopposed exploitation"); and finally "by the white man's most potent weapon: booze" (which is "the enemy's delight").

Puxley described the way to true liberation by quoting from the guidelines of the Tanganyika African National Union. (Tanganyika later became Tanzania.) The objective of one TANU guideline he quotes is "to make the people aware of our national enemies and the struggle they employ to subvert our policies, our independence, our economy and our culture." Another TANU guideline quoted by Puxley states: "The greatest aim of the African revolution is to liberate the African. This liberation... is achieved by combatting exploitation, colonialism and imperialism." Puxley asserted that "just as the liberation of the African is the primary goal of the African revolution, our greatest aim is the liberation of Indian people."

"There is no better way to achieve liberation and resist exploitation than for the people to become aware of their situation and understand the nature of the forces which oppose them," Puxley continued in his paper. "The role of the organizations from this perspective is that of a spear-head of awareness, working with people at the community level, discussing situations familiar to community people, and helping them to develop a deeper understanding from which they can take action, themselves, to improve their condition."

A two-day workshop of representatives from the Brotherhood and Metis boards, held at Camp Antler near Yellowknife early in July 1975, ended the efforts of the two organizations to develop a common proposal for the settlement of native land claims and escalated the problems facing Wah-shee. After the conference, the Metis Association and the Brotherhood each produced different versions of what had been agreed to. Rick Hardy claims that the Brotherhood version was changed after the Indian representatives had consulted with their white advisors, who were not at the meeting.

On the second day of the workshop, "it ended up that Wah-shee and I were taking on the rest of the people there," Rick later told

186

me. "Our point of view was that we should be getting involved in the day-to-day activities that would...put bread and butter on the table...The concept was to start working towards helping people to better their lives physically. The other side felt that we should be concerned more with the overall philosophical questions and establishing the new Socialist state or whatever it was they wanted. I believe it was from that particular session that the opposition decided that Wah-shee had to go."

With the two organizations no longer working together on a joint land claims proposal, government funding was temporarily cut off and plans for a joint assembly to ratify a common approach were scrapped. With the aid of $100,000 from the Anglican and Roman Catholic churches, the Brotherhood proceeded with its own conference, and at Fort Simpson on July 20 adopted its famous Dene Declaration.

Some of the philosophy that lay behind the Dene Declaration is contained in a prior paper prepared at the Brotherhood offices and marked "Draft Dene Declaration, for discussion purposes only."

"Real power," claimed the draft, "lies with a handful of large companies who operate with the full co-operation of both the Territorial and federal governments," the same companies that exploit and oppress the people of the Third World countries. "The great majority of people in Canada are like ourselves in being relatively powerless in the face of big companies and governments ...By joining us in our struggle people can begin as well to liberate themselves."

"We have been made Canadians by decree and not by our free choice," the paper states. "Understandably our first choice would be to be once again a sovereign people. But we are realistic and know the white man is powerful."

The conformity required for a collectivist society was made explicit. "Some of our people who are becoming involved with industry and government are slowly coming to believe in their ways and are forgetting where they came from," the paper warned. "They need clear direction. We know that there are powerful forces arranged against us. That is why we have not hesitated to appeal to others to support us in our just struggle."

At the same time, the enemies must be dealt with:

Our struggle is like a war, but a peaceful one. On each front there is an enemy. On each front there are allies. On the external front the enemy is those not a part of the Dene Nation who resist and deny the achievement of our goals.... There are also Dene who are the enemy. There are Dene who would betray and are betraying their brothers in the struggle for their goals. These are the Dene who work for the enemy against their brothers. These are traitors to the cause of the Dene Nation. We must learn to identify such persons... There are Dene who have not yet learned who the real enemy is. It is the duty and responsibility of Dene who have learned to recognize the real enemy to educate their brothers.

The short Dene Declaration Statement of Rights, adopted by the Fort Simpson Assembly, was purged of the worst extremes in the earlier draft paper. The final Declaration states, in part:

We the Dene of the NWT insist on the right to be regarded by ourselves and the world as a nation. Our struggle is for the recognition of the Dene Nation by the Government and people of Canada and the peoples and governments of the world. As once Europe was the exclusive homeland of the European peoples, Africa the exclusive homeland of the African peoples, the New World, North and South America, was the exclusive homeland of aboriginal peoples of the New World, the Amerindian and the Inuit... No where in the New World have the native peoples won the right to self-determination and the right to recognition by the world as distinct people and as nations...
The challenge to the Dene and the world is to find the way for the recognition of the Dene Nation. Our plea to the world is to help us in our struggle to find a place in the world community where we can exercise our right to self-determination as a distinct people and as a nation. What we seek then is independence and self-determination within the country of Canada. This is what we mean when we call for a just land settlement for the Dene Nation.

Less than four months later, a revolt by the staff and consultants succeeded in expelling James Wah-shee as president of the Brotherhood.

In a news release issued October 31, 1975 the Brotherhood announced that the board "has lost confidence in James Wah-shee as president," that a resolution to remove him from office would be brought forward at a general assembly to be held in Aklavik

188

(December 1–5), and that in the interim vice-president Richard Nerysoo had been named chief executive officer.

Wah-shee responded with his own news release, accusing the staff and white consultants of setting Brotherhood policy and causing his removal. "It was never meant to be that the advisors or staff should be deciding the direction of the Brotherhood," Wah-shee stated. "It is not the job of staff and advisors to play any part in the political decision-making of the Indian Brotherhood, yet I know that in this particular situation, the staff and advisors have played a major part."

Five people comprised the Brotherhood's board, Wah-shee and four band council chiefs. One of the board members was chief Alex Arrowmaker of Fort Rea, the home community of both Wah-shee and Erasmus, then staff member in charge of community development. In a signed statement published six days later and never denied, Arrowmaker gave his version of what had happened:

"The problem we are talking about has come from the white people who got together with other staff in the Brotherhood's office," Arrowmaker stated. "They met as a group and then came into a workshop three members of the board were having with James Wah-shee and his vice-president. The staff ... said they refused to work with James Wah-shee as their president and that if the board did not get rid of him that they would all quit." That was on October 30, and Arrowmaker was not present. He arrived in Yellowknife the next day and said that he "saw that the staff and white people were acting like judges in a court and they were saying he [Wah-shee] had to be thrown out ... George Erasmus did all the talking and the white consultants handed him all the notes on what to say." Three members of the board—with Arrowmaker and Wah-shee absent—met later that day and agreed to remove Wah-shee from office.

Wah-shee later resigned his seat on the NWT Legislative Assembly. In his farewell address, on May 28, 1976, he bitterly denounced his removal from office, urged "a just and speedy settlement of the land claims of native people," and criticized the Brotherhood for spurning the legislative process. He had been, he claimed, "betrayed and cast aside" by the "treachery, the conspiratorial tactics" of "the Brotherhood's revolutionary council."

"The position adopted by some native leaders and their white advisors that it is politic to ignore and deprecate the political institutions that have so much effect on our daily lives is both foolish and immature," Wah-shee stated. He said that the laws "apply to, and protect, the interests of all people, including the Dene. Native people must be present, be heard, and more important, be counted when the laws are made that so affect our lives and livelihoods. To turn our backs on such institutions is unrealistic and irresponsible."

Wah-shee had vowed that he would fight for re-election as president of the Brotherhood, but in the end he did not. At the Brotherhood's assembly that June, George Erasmus was elected president. Wah-shee, reported the *Saint John's Calgary Report*, was disconsolate and discredited, separated from his wife and two adopted children, and had been fined a hundred dollars for possession of a small amount of marijuana.

The Brotherhood's concept of a Dene Nation was further defined during the following months in presentations to the Berger Inquiry and the National Energy Board by Erasmus, by Brotherhood consultants, and in speeches by Erasmus.

Economist Mel Watkins, who testified before both Berger and the NEB, said that the central purpose of the native claims is to end the "terrible historic process" by which "industrial class-capitalist Canada" has separated Indians from their lands and livelihoods, with the lands monopolized by "a single social class" leaving Indians "permanent members of the underclass."

Watkins said the Dene seek "alternative community-based economic development under their control." The way to achieve this is with a "two sector economy" in the north, including "a non-renewable resource sector under white ownership but subject to Dene control, and a renewable resource sector under Dene ownership and control." Revenues from the non-renewable resources would subsidize the renewable resource activities. "The right to alternate development must include the right to tax the non-renewable resource sector, or to impose royalties thereon, so as to fund the Dene economy and the Dene institutions, which will permit continuing Dene development."

Peter Russell, professor of political economy at the University of Toronto, constitutional advisor for the Dene, and co-chairman of the Dene's Southern Support Group, defined for both Berger and the Energy Board, what was meant by the Dene Nation. Not a separate sovereign nation, he said, but "a distinct cultural entity...

190

a distinct nation or national group within the Canadian state." This, he said, would accord with "the spirit of confederation." He claimed that "with its diffusion of government authority and its lack of ethnic homogeneity" the Canadian Confederation "has not been an easy one in which to govern, but it may well be the most liberal mode of self-government for a large continental nation-state that the world has known."

Donald Simpson, the other co-chairman of the Dene Southern Support Group, a professor of the history of education at the University of Western Ontario, on leave to the International Development Research Centre, spoke at the NEB hearings about his sixteen years of experience with international aid and development agencies. "I have learned a greal deal about colonialism as it was practised both in Asia and Aklavik," he said. "A good deal of it is a sad story of man's infinite capacity for greed."

Native people must establish a sense of self identity and self-worth before they can improve their circumstances, according to Simpson. This can be established only if they control the investments which affect their lives. "If they do not control those, they do not control the technical configuration of those investments; and if they do not control that, they do not control a large proportion of the conditions which determine the nature of their society. And, if they do not control that proportion, they might as well just forget about all the other things they might do to preserve and develop their own social and cultural identity and integrity."

Puxley also testified before Berger, presenting a philosophical dissertation on the nature of colonialism, condemning "the corporations whose imperatives define our choices," and "our technological and advanced capitalist organizations."

Puxley placed great emphasis on education. He called it a colonial tool which "has come to condition man, to govern him and to deny the very human process of deconditioning...The education system of the Northwest Territories blandly continues to serve but one purpose, the maintenance of a colonial and dehumanizing experience for the Dene...Those who at present are nominally charged with the responsibility for Dene education are totally incapable of carrying out the task...since they have not even addressed the problem of colonialism in their own lives."

Erasmus, too, in testimony before the Energy Board, stressed the importance of education. "We have been colonized," he said.

"What we need are institutions that will help us decolonize, and the educational system is a very powerful tool."

The Dene, Erasmus told the board, want the right to determine who could vote in the proposed Dene Nation. "We want to have a separate jurisdiction to be able to protect ourselves from an invasion of people coming up. We would want the power of setting residency requirements and who could vote...we might decide that non-Dene people coming into our area would have to be able to speak a Dene language."

In July, 1977, Erasmus presented to Indian Affairs minister Hugh Faulkner the Dene proposal to split the Northwest Territories into three separate territories. In one territory, the majority population would be Dene; in the second, Inuit, and in the third, non-Dene. Each territory would have its own government with powers similar to those of a provincial government. "Each territory would set up a legislature according to the democratic decision of its respective populace," the proposal states. "The Dene and Inuit would institute traditional native forms of government." To deal with matters of common interest, there would be an over-all Metro or United Nations form of government, to which each of the three territories would send representatives.

"As long as the recognized form of government in the NWT is a foreign institution in which native people do not choose to participate, decisions made by that government will continue to be oppressive," Erasmus asserted in a statement accompanying the proposal. (In fact, however, native people do participate in the present form of government in the Northwest Territories. They turn out in large numbers to vote; they comprise a majority of the council members in most of the settlements, as well as a majority in the NWT Legislative Assembly, and have elected a native as Member of Parliament for the NWT.)

The right to "self-determination" sought by the Dene was strongly supported by Judge Berger in his report. In a written critique of the Berger report, W. H. McConnell, professor of law at the University of Saskatchewan, observes: "It is possible that the establishment of self-governing native political units in the North might be the first stage in a process in which they ultimately claimed full independence, assisted by allies in the U.N." and from Third World countries.

McConnell quotes Professor Sam Deloria, an Oglala Sioux who

is a law teacher at the University of New Mexico and, McConnell says, an associate of Erasmus: "It is one thing for American domestic courts to hold that they have no constitutional authority to compel Congress or the Executive to respect the sovereignty and international personality of Indian tribes; it is quite another for the international community to consider itself bound by the same constitutional branch."

McConnell suggests that both Deloria and Erasmus,

> tend to regard the self-determination of native peoples as a global phenomenon through which widely dispersed native minorities will attain local autonomy as a halfway house to ultimate sovereignty and independence. To say this is not to impugn their honesty or sincerity, for their immediate aim is merely greater autonomy for native peoples within their respective federal systems. However, the very articulation of their immediate goals in terms of such an international norm as "self-determination" tends to suggest . . . that their ultimate identity and the very terms in which their leaders state their case indicate that there are strong separatist currents underlying the rhetoric of their arguments. It is the native elite which puts forward such arguments.

The suggestion that the Dene leaders might seek the help of other countries in their quest for self-determination, or independence, is echoed in the Dene statements.

The Dene Declaration itself pleads "to the world . . . to help us in our struggle to find a place in the world community where we can exercise our right to self-determination as a distinct people and as a nation."

In a statement presented to then Indian Affairs Minister Warren Allmand in Ottawa, October, 1976, Erasmus declared that "our demand for the right to self-determination is supported in international law." He quoted from the UN International Covenant on Economic, Social and Cultural Rights, adopted in 1966: "All peoples have the right to self-determination. By virtue of that right they freely determine their political status and freely pursue their economic, social and cultural development."

In Geneva (September, 1977), Erasmus addressed a UN conference on Discrimination Against Indigenous Populations. He urged support for the quest for "international recognition of the right to self-determination of aboriginal minorities in independent countries. Eminent scholars . . . have argued in support of aboriginal nations, such as the Dene, asserting their right of self-determination as a legal right under public international law. The world community must aggressively declare their recognition of that fact. Only the weight of international opinion may force governments and majorities to abandon their resistance to the vital change in human relations being initiated by the aboriginal nations."

The pay-off for the white consultants who had provided the ideological base for the Brotherhood, had ousted James Wah-shee because he was no longer a true believer, and had helped put Erasmus in charge, came late in November, 1977. Except for Mel Watkins, who had already returned to the University of Toronto, Erasmus fired them all. The following April, the name of the organization was changed to the Dene Nation. A renewed drive was launched to recruit Metis members who agree with the concepts of the Dene Nation. There was no indication that the goals or direction of the organization had changed. The white advisors may have left, but their philosophy remained.

No matter what might be said about the concepts of the Brotherhood and the Dene Nation, it had made an important start on at least one essential task—providing native peoples with a sense of identity and self-worth. Don Simpson was right, that without this no significant improvement in the circumstances of native people is likely. But just as essential will be the economic and career opportunities to meet the rising aspirations of young native people. Without this, frustrations, resentment and tensions may grow like a cancer in the land.

THE TANZANIAN MODEL

"Before the coming of the non-Dene," Erasmus has said, "the Dene completely ruled their own lands, resources and themselves through a democratic government of their own. Our people had a full, complete culture, including education system, social institutions, religious ideology, justice system, and economy based on self-reliance and stressing a political system based on personal responsibility for the collective interest."

What the leaders of the Dene Nation now propose is a return to this traditional native society and economy, with community-owned enterprises, but modernized by the use of the white man's technology of snowmobiles, rifles and outboard motors.

But in fact, says Rick Hardy, Tanzania is the model for the Dene Nation. The Dene's internal papers are laced heavily with references to the writing of Tanzanian president Julius Nyerere, considered the father of African socialism.

A description of the community enterprises being developed in Tanzania sounds strikingly like what the Dene leaders seem to have in mind. In the Tanzanian village of Luhanga, reports *Time Magazine* (March 13, 1978), "the volunteer village militia combats crime, the village-owned dispensary and clinic combat disease, the village-owned furniture shop and tinsmith combat unemployment. A woman's cooperative sells milk and soft drinks, while profits from the village's enterprises fund a school and day-care centre ... members are encouraged to work in the communal enterprises. Instead of pay, they receive 'points,' which entitle them to a share of the profits from the village's communal projects."

The country was once the German colony of East Africa. In 1954, Nyerere was instrumental in organizing the Tanganyika African National Union (TANU) leading to the establishment of the independent Republic of Tanganyika in 1961, followed by a merger in 1964 with Zanzibar to form the United Republic of Tanzania. It is a nation of some fifteen million people and one hundred and twenty-six tribes.

Nyerere calles his political philosophy "ujamaa," meaning familyhood. He says that the basic elements are "mutual respect, sharing and work."

The concepts of the Dene Nation echo the writings of Nyerere:

"The biggest crime of oppresion and foreign domination in Tanganyika and elsewhere, is the psychological effect it has on the people who experience it. A vital task for any liberation movement must therefore be to restore the people's self-confidence, and it was quite clear to us that a multi-racial TANU could never do that ... Only by creating and developing our own exclusive organizations could we begin to develop confidence in our own abilities ... For these reasons TANU became a racial organization," but based on "racial equality."

The ideal society, Nyerere has written, requires "the group's joint ownership of basic property. It is, and must be, 'our' house, 'our' food, 'our' land, for only under these conditions can equality exist among the members." In *Ujamaa*, he wrote: "To us in Africa, land was always recognized as belonging to the community." Man's right to the land "was the right to use it; he had no other right to it, nor did it occur to him to try and claim one... The foreigner introduced a completely different concept, the concept of land as a marketable commodity."

If the Dene's concepts are not quite entirely indigenous to the Mackenzie Valley, neither are Nyerere's concepts of Ujamaa entirely indigenous to Tanganyika.

Julius Kambargage Nyerere is the son of a minor tribal chief who had eleven wives, and Julius was one of the few sons chosen to attend school. He was a brilliant scholar, and in 1949 went to Scotland to attend the University of Edinburgh for three years where he studied history, economics and philosophy. "My ideas of politics were formed entirely during that time," he has said. (Not, obviously, from the writing of that Scottish professor of moral philosophy, Adam Smith.)

In seventeen years of independence under the leadership of Nyerere, Tanzania had not yet managed to attain utopia. In 1977, Tanzania's per capita GNP and "physical quality of life index" (as measured by the Overseas Development Council in Washington) ranked below such countries as Kenya, El Salvador, Rhodesia, the Sudan, Morocco, New Guinea and Bangledesh.

Democracy in Tanzania is of the one-party variety. Nyerere has received more than ninety per cent of the votes at elections, but there is no one else to vote for. The Manhattan-based Freedom House rates the "political freedom index" in Tanzania below other low-rated countries like South Africa, Zambia, Rhodesia, Indonesia, Ecuador, and on a par with Chile.

Like the leaders of the Dene Nation, Nyerere has stressed the importance of controlling education. He has argued that in order to preserve the principles of ujamaa socialism, "the whole of the new modern education system must also be directed towards inculcating them. They must underlie all of the things taught in the schools, all of the things broadcast on the radio, all the things written in the press. And if they are to form the basis on which society operates, then no advocacy or opposition to these principles can be allowed."

A true Marxist version of democracy, and an inspiring model for the Dene.

196

THE HEARINGS

The accusation has been made that the views of native peoples presented to the Berger hearings did not reflect their real attitudes, that their evidence had been stage-managed by the native organizations, that they had been pressured into paroting the party line of the Brotherhood. Michael Jackson, special counsel to the Inquiry in charge of arranging the community hearings at which native people spoke, says that there have been allegations that he participated in stage-managing these presentations.

Berger deals with these accusations in the second volume of his report: "Such allegations reflect a lingering reluctance to take the views of native people seriously when they conflict with our own notions of what is in their interest. Such allegations, advanced in order to discredit the leaders of the native organizations, lose their force when measured against the evidence of band chiefs and band councillors from every community in the Mackenzie Valley and western Arctic, and against the evidence of the hundreds of native people who spoke to the Inquiry."

Michael Jackson is a professor of law at the University of British Columbia. Before joining the Berger Inquiry, he had acted for five years as legal advisor to a number of Indian groups on the west coast and had taught a course on native rights for four years. After the Berger hearings, he joined George Erasmus on speaking platforms advocating the cause of the Dene Nation. He speaks with a clipped Oxford accent. With a lean face, a tanned complexion, a long drooping mustache, hair that neatly cascades down the length of his back, he would make a perfect cast for the movie role of Grey Owl.

Jackson, in an interview early in 1978, also denied the allegations: "Some of the attacks made on the community hearings are particularly troublesome." He referred to "the allegations which were made, sometimes directed towards me, sometimes towards the Brotherhood's white advisors, that the community hearings were always stage managed." That this was not so, he said, would be clear to,

> anyone who had any understanding of how the ground work for these hearings was laid and the extent to which the communities were really told that the hearings were their's ... the hearings were really up to the people in the villages to determine, both the context and structure ... it was the process least

197

capable of being manipulated by outsiders. If the Brotherhood wanted to stage manage the hearings they would have had a very rigid kind of procedure in which just the chiefs spoke... The less people that spoke the more likely you could get a unanimous, uniform line... To the extent that the hearings becomes a totally open process in which not only everyone is encouraged but takes the opportunity to speak, it's very difficult to stage manage a whole village.

He said that native people are "very individualistic" in terms of individual participation in group decision-making. "Those villages would be the most difficult kinds of communities to manipulate."

There is no question that the views which native people expressed did, in fact, reflect their real attitudes and, often, very deep convictions. But there are other questions that are not resolved.

To what extent did those views support a ten year deferral of a pipeline? To say that land claims must be settled first is not necessarily to support a long moratorium, a distinction which has not always been made. To what extent were the attitudes of native people influenced by the Marxist anti-resource development philosophy supplied by the Brotherhood to its field workers in books, films, posters, pamphlets and workshops? Why was there not one question to any of the hundreds of native people who testified about what information they had on which to base their conclusions?

Certainly, Berger did not hesitate to express his scepticism about some of the evidence which tended to conflict with the case for the Dene Nation. In his report, he characterized estimates presented by the NWT government as "official willingness to justify the construction of the pipeline on the basis of an inflated figure for unemployment." As for other evidence, he wrote that "consultants who now recommend the construction of the pipeline on the grounds that it will benefit the native people of the north, will be succeeded by consultants willing to support whatever conclusions government and industry are than anxious to justify."

"We sought to avoid turning the Inquiry into an exclusive forum for lawyers and experts," Berger wrote in his report. To accomplish this, the hearings were divided into two phases: formal hearings in Yellowknife where the evidence was subject to cross examination, and more informal hearings at thirty-five communities in the Mackenzie Valley and Delta region, as well as in ten cities across southern Canada. Close to a thousand residents testified at the northern community hearings.

Jackson said that his job was to go to the communities and explain what the hearings were all about—that they were not just about a pipeline but "about the future shape of northern Canada, and the evolution of political institutions, the evolution of native and white self-determination and a whole range of other issues which were put into perspective by the pipeline," and "to help the people in the communities prepare for the hearings." He was aided in this, he said, by field representatives from the native organizations, as well as by representatives from the municipalities and chambers of commerce.

Jackson said further that the native people "were very well informed, not necessarily as to the specifics of the project... but as to the kind of impact which the project would have on their lives, on the economic choices open to them, on the possibilities of the north being a place in which native people were the prime movers, as opposed to being the moved."

How did the native people get their information about the pipeline?

"In terms of the pipeline," Jackson said, "the Indian Brotherhood set up a network of field workers which took to the villages. The Committee for Original Peoples Entitlement [COPE] did the same thing. They held workshops for their field workers, provided them with the information received from the companies.... Because we had a year from the time the Inquiry was set up until the hearings there was a year's lead time in which people could be provided information."

The Brotherhood, however, provided their people with very little information from the pipeline companies for one very simple reason: the Brotherhood for the most part declined to talk with the pipeline companies. A very few meetings had been held when Wahshee was president, but then he was accused of being co-opted by the enemy and removed from office. There was little opportunity for native people to gain an understanding from the Brotherhood about what the pipeline involved, other than the Brotherhood's view of what the pipeline involved.

While virtually every native person demanded a settlement of their land claims first, the views about whether, or when, a pipeline should follow were mixed, or often simply never stated. The position of the Metis Association was spelled out by Rick Hardy in final argument on the last day of the Inquiry hearings:

"First, we are of the view that a settlement should be made between the native people of the Northwest Territories and the federal government before construction of the Mackenzie Valley gas pipeline," Hardy said. "Our second point," he added, "is that only a minimal number of years are necessary from the time a settlement is made with native people to the time when a pipeline is started." Hardy said that Metis people see their "future in the melieu or context of a pipeline or other development of the north.... We have come to depend on wage employment and many Metis people have small businesses. We cannot now endorse or suggest an economic future which will in any way hinder or adversely effect such an economic state. We therefore look to the construction of the pipeline as one of the major economic projects which we wish to take part in ... if we as Metis people are to survive and continue to grow socially and culturally we must first be economically and otherwise secure."

"I'm tired of people telling me they are anti-pipeline, and not giving me any reason why they are anti-pipeline, and insist that I should be so likewise," Mrs. Jeanette Ross, a Metis, told the Inquiry at Norman Wells.

"Yes, sir, I've been pressured into saying I'm anti-pipeline," Mrs. Ross asserted. "So is the rest of the Metis people around here, probably; and I often wonder if probably the other settlement leaders do the same."

No one asked Mrs. Ross whom she was pressured by; how she was pressured, why she was pressured, or when. The accusation just lies there on the pages of the hearing transcript, unrefuted.

Charlie Furlong, an Indian in Aklavik, was one of the field workers employed by the native organizations to provide information to the communities prior to the hearings. Two months after the first volume of the Berger report was released, Furlong talked about this in an interview recorded for a film.

Well, the government financed the native organizations of the north to prepare for the Berger hearings. Prior to each community hearing, field workers of the organizations came into the community, and I myself worked for those organizations, and people of the communities were given all the negative parts of the pipeline and no real information was given concerning how a pipeline was being built. So I think fear was put into a lot of community people concerning the pipeline. So I think this is why, throughout the whole Northwest Territories, every community sort of patterned out like Aklavik did, the first day of the hearing. Everybody was against the building of a pipeline and so from there on right down the Mackenzie, all the communities said the same thing: no pipeline until land settlement. And this was, we were told this by the Brotherhood and the Metis Association before the communities and so I think really field workers put fear into people concerning the pipeline . . . this is the message that the field workers put across in all the communities and I think the Brotherhood and Metis Association did a damn good job of it. Now I think some people are sorry that it's been done.

THE REPORT

Judge Berger's report on the pipeline Inquiry pins the future hopes for native people in the Mackenzie Valley and Delta on an idealized vision of a racially-segregated native society based on the harvesting of wildlife and community-owned enterprises, as articulated by advocates of the Dene Nation.

This would require, Berger admits, public subsidy. How much, he does not estimate.

Even with unlimited public funds, there is no assurance that a society and economy of this nature can be attained that is viable enough to meet the requirements of a population which will increase fifty per cent in the decade to 1985.

Nor is it clear that this vision of a collectivist utopia, even if it could be attained, is an aspiration shared by the majority of the native people of the region.

Berger did not reject a Mackenzie Valley pipeline because it would fail to provide significant employment for native people. It was, in fact, the very prospect that more native people would be drawn into the wage economy which was his greatest concern.

Implicit in Berger's analysis is the view that native people cannot cope with wage employment, that for the majority of native people it can result only in increased social pathologies: alcoholism, crime, social disorders, violent deaths.

The conflict between the native economy based on renewable

resources and the wage economy based on non-renewable resources is "irreconcilable", Berger states.

Given this assessment it is impossible, in Berger's view, to contemplate concurrent development of both sectors of the northern economy in order to provide native people with maximum individual choice and opportunity. The native economy must be developed first, during the ten years of the moratorium, in the hope that this will reduce the number that will be attracted to the wage economy. Failure to develop the native economy first, Berger argues, would force native people into the wage economy. There is no admission that a decade of freeze in the wage economy might force native people into continued hunting and trapping pursuits. This, says Berger, is what they want.

What happens if the native economy cannot expand fast enough to meet the needs of a population that will increase by half during the decade of the wage employment freeze? That is one of the perils which now confronts the native people in the Mackenzie Valley and Delta region.

Berger estimates the total 1974 population of the Mackenzie Valley and Delta at some thirty thousand, of which approximately half are native people. The native population, he further estimates, includes seventy-five hundred Indians, twenty-three hundred Inuit and forty-five hundred Metis and non-status Indians. Other estimates of the Metis population have ranged from a low of twenty-five hundred by the Brotherhood, to a high of twelve thousand by the NWT Government.

The birth rate of native people in this area during the early sixties was among the highest in the world, five times the average for the rest of Canada. Although it has since declined, it was still nearly three times the national average in 1974. At that time, half the native population in the area were under fifteen years of age, and less than eleven per cent were over fifty.

Can the wildlife harvest be increased enough to sustain this growing population?

Berger estimates that trapping of furs could be increased several fold without endangering the population of fur-bearing animals. But there are problems. It would require trapping, he says, in more remote and under-exploited areas, at greater costs. Already trappers and hunters do not receive enough cash income to cover their expenses for securing furs and the game food they use (an average

202

two hundred and forty pounds per person per year). Hunting and trapping expenditures for the five-year period to 1975, Berger reports, have averaged $3.5 million per year, compared with cash income of only $1.2 million from furs plus a substantial portion of the food consumed by the hunters and trappers and their families. The cash expenditures, Berger admits, are "substantially more than the most optimistic potential yield" could generate in cash income. That is why, he says, native people have had to take seasonal wage employment with the oil companies.

The native economy advocated by the Dene leaders and by Berger will require subsidization, Berger admits, on two levels: to meet the shortfall in the annual expenditures involved in harvesting wildlife, and to fund the proposed community-owned enterprises (whatever they might be). "Financing at this higher level," Berger says, "will have to be more generous than it has been in the past." He claims that this can be achieved "with a comparatively small capital outlay. A reasonable share of the royalties from existing industries based on non-renewable resources in the Mackenzie Valley and the western Arctic would suffice."

But royalties collected by the federal government on all mineral production in the Northwest Territories are less than two per cent of what the federal government already spends to subsidize the NWT economy. In 1975, the federal government collected approximately $4 million in royalties on NWT mineral production. Federal expenditures in the north, in 1977, exceeded $300 million or $8,500 for every NWT resident. (Largest federal spending in the north is through the NWT Government. For the 1976-77 fiscal year, southern taxpayers provided eighty-seven per cent of the budget of the NWT government. Other federal spending in the north is largely by the Department of Indian Affairs and Northern Development, as well as recently instituted DREE programs.)

Berger blasts what he describes as the "huge subsidies ... provided to the non-renewable resource industries," ignoring that whatever subsidies these industries receive are but a fraction of the taxes and royalties they pay. The native economy, meanwhile, is not only to be subsidized, but exempt from taxes and any burden of the cost of government services and infrastructure. Suggestions that "a renewable resource economy should be expected to generate profits that can be taxed to support government services and infrastructure ... are clearly at odds with native peoples' objectives and their economic future," Berger states. If such suggestions "continue to prevail then an economy based on renewable resources that so many native people desire has no viable future, and native people will never have a real chance."

The cost to southern taxpayers of shutting in Delta gas reserves for ten years while trying to establish a subsidized native economy on a limited resource base will be great. With the rapidly growing population and unemployed labour force, present yearly federal government spending of $300 million will have to be increased. It is difficult to see how it could be held to less than five billion dollars during the ten year period. At the same time, the locked in Delta gas reserves will represent a loss of two to three billion dollars in the public sector during the same decade.

Berger remains undaunted by past failures in the north. "Northerners ... are highly conscious of the failures of many small-scale projects or enterprises based on renewable resources," he reports. "Almost all of them have been associated with government, which to many is explanation enough." How it will be possible to rely on government subsidy and avoid government involvement is not explained.

Berger subscribes to the theory advanced by Watkins that resources in the hinterland are exploited for the benefit of the southern metropolis with little gain to the region from whence they came. "Large-scale frontier projects tend to enrich the metropolis, not the communities on the frontier," he asserts. "The mining industry has ... taken its profits out of the Northwest Territories, and the oil and gas industry will do the same." He argues that, "it is an illusion to believe that the pipeline will solve the economic problems of the north. Its whole purpose is to deliver northern gas to homes and industries in the south."

Resource development is thus seen as a zero-sum game: if someone wins, someone has to lose. If this were true, it would be only Ontario and other metropolitan centres that would benefit from Alberta's oil and gas, since the whole purpose of the pipelines from Alberta is to supply oil and gas to other areas. In fact, Alberta has possibly the most prosperous economy in North America.

In addition to an expansion of the wildlife harvest, Berger also envisions the native economy developing secondary activities based on renewable resources. As an example, he suggests a "complex of small-scale enterprises, based on logging, the production of logs and lumber, and the construction of houses." Berger admits that, with the small timber resources of the north, this would be more costly than importing prefabricated modules; however, "the total benefits to the north in terms of local enterprise, training, employment and income would be far greater." He does not say who would pay for this extra cost, nor how many jobs this would provide. In 1978, a NWT government representative said that even if the economic problems could be overcome, the limited northern forestry resources could provide fewer than five hundred jobs. Meanwhile, the labour force is growing at a rate of one thousand every year.

"How many people can the land ultimately support even when the renewable resources of the north are fully utilized?" Berger asks. The question is never answered.

In the first volume of his report, Berger states that, "The north is, in fact, a region of limited biological productivity. Its renewable resources will not support a large population." In the second volume of his report, he concluded: "Perhaps some groups of native people who live in areas that are relatively poor in renewable resources, cannot hope to increase their production of fur or food in any significant way."

Berger found that "a healthy economy based on renewable resources offers employment far beyond the primary production." He did not acknowledge that, in similar fashion, a pipeline would provide employment beyond the two hundred and fifty jobs involved in operating the line itself. There was no recognition of pipeline-related jobs in exploration, development, production, gas processing plants and transportation and other services. The National Energy Board acknowledged that these activities would provide twelve hundred long term seasonal jobs and two thousand long-term year-round jobs. Even the two hundred and fifty pipeline jobs, Berger concluded, would be "Jobs of a technical nature and will have to be filled by qualified personnel from the south." The fact that northern native people had already been employed in a pipeline and petroleum industry training program for more than six years was deprecated by Berger.

Berger argues that native people cannot cope and compete in wage employment, do not want it, and that it produces disastrous social effects, thus the fewer native people that work on a pipeline, or pipeline related jobs, the better.

He says that "the real danger" of large industrial projects "will not be their continued failure to provide employment to the native people, but the highly intrusive effects they may have on native society and the native economy." Seasonal jobs are good only if the wages are used to buy hunting and trapping supplies. Wages "used to buy provisions and equipment, such as snowmobiles, guns and traps...serve to re-inforce the native economy and culture. But much of the cash that is earned is not so used, and this has had consequences that have been destructive and divisive."

"All the evidence indicates that increases in industrial wage employment and disposable income among native people in the north brings with it a dramatic increase in violent death and injuries," Berger stated. He blames wage employment as the cause of these social pathologies: high rates of violent deaths (nearly two and a half times the national average), excessive alcoholic consumption (three and a half times the national average), crime, and other social disorders. "I am persuaded that the incidence of these disorders is clearly bound up with the rapid expansion of the industrial system and with its persistent intrusion into every part of the native people's lives."

In the second volume of his report, Berger concludes: "If...the development of renewable resources is given priority, then the supply of local northern labour available for pipeline work might decrease considerably. That result is greatly to be wished for: my thesis throughout has been that we should create the conditions that would enable the native people to strengthen the renewable resource economy in the north and thereby reduce their vulnerability to the social and economic stresses that industrial employment has, in the past, visited upon them."

But many others—particularly in the NWT government—have persuasively argued that the social pathologies will increase unless there is an expansion of opportunities in both the wage sector and the traditional economy. They argue that the traditional economy cannot absorb the exploding population, that many young native people aspire to rewarding wage employment, and that a lack of opportunities will produce despair, bitterness, frustration, and further social pathologies.

Berger asserts that native people prefer traditional occupations over wage employment, although the hearing record on this was contradictory. Dr. Stabler of the University of Saskatchewan accuses Berger of ignoring this contradictory evidence and selectively using other evidence to support his position. Stabler cites reference in Berger's report to a study for the federal government by Derek Smith, published in 1975. "Judge Berger quotes Smith in support of the 'importance of the land in the native culture' theme," Stabler writes. "Yet the major conclusion of Smith's survey directly contradicts the point that Judge Berger is attempting to establish."

Smith's paper reported the results of a survey among native students in grades seven to twelve at Aklavik, Inuvik and Fort McPherson, who were asked to rank forty-eight occupations according to their preferences. The jobs ranked tops by the native students were, in order, pilot, stewardess, radio operator, nurse and doctor. The four land-based jobs were game officer, which ranked twentieth in position; fur garment worker, thirty-third; hunter-trapper, fortieth and reindeer herder, forty-seventh just ahead of garbageman. Native students, Smith concluded, "esteem professional and skilled jobs, urban-type working conditions, and reject the seasonal, unskilled rural and outdoor sorts of occupations with which native people have so consistently been identified in the past."

Berger points to the lack of success of native people in competing for jobs, and blames this, in part, on the education system, which he says is alien to native students. He notes that of nineteen hundred federal government employees in the Northwest Territories in 1976, only two hundred and fifty were native people, while in the Territorial Government, only six hundred and three or twenty per cent of its employees were native people.

Less than fifteen per cent of young northern native people had received any formal schooling by 1950, while by the end of the sixties, between ninety and ninety-five per cent of the school-age children were attending school. Berger cites one study which claims that in 1970 only seven per cent of the native people in the Mackenzie Valley completed high school and none held a university degree.

Berger says the educational system for northern native people will not succeed unless they control it themselves. He claims that the policy of the NWT Government "to transfer responsibility to the local communities, to make the curriculum culturally relevant, and to train native teachers," will not work. "The reason is simple," he says: "one people cannot run another people's schools."

But Berger is too pessimistic, too paternalistic. Native people are graduating from schools, holding more of the available jobs in the north, and assuming an increasingly greater role in northern society. Where Berger reported that only twenty per cent of the NWT Government employees were native people in 1976, by 1978 this had risen to nearly twenty-seven per cent (seven hundred and twenty-eight out of a total of twenty-seven hundred and forty). Probably fewer than two hundred native NWT residents had completed high school by 1969, but in 1977 alone ninety-three native students graduated from NWT high school. While the report cited by Berger claims that no northern natives had graduated from university by 1970, off hand I can think of four—from the limited number of northern natives I know—who have. There's Noah Carpenter, an Inuit from Sachs Harbor and a medical doctor, who testified before Berger in Edmonton. There's Addy Tobac, in Inuvik, formerly one of the Brotherhood leaders. Two others are personal friends: Nick Sibbeston a Metis, who has his own law practice in Yellowknife, and Frank Hanson, an Inuit engineer who now has his own prosperous business in Inuvik.

Stabler characterizes Berger's report as "an articulate but emotion-based argument against the pipeline proposal" which offers "rhetoric . . . where facts are required." He concludes that Berger has created a situation in which a satisfactory resolution may be impossible, and states:

> If the pipeline now is built it will appear to the native people that their wishes have been ignored. If, instead, the proposal for a moratorium is accepted and it subsequently proves impossible to create a viable economy around the renewable resource sector, then rising unemployment, unrealized expectations, and a growing dependence on transfer payments will follow. In either case the disillusionment of the native people is a very real possibility. This disillusionment may in the long run be more detrimental to the native people than that which Judge Berger envisages from the construction of the pipeline.

THE ALTERNATIVES

"Perhaps a redefinition of the relationship between the Government of Canada and the native people can be worked out in the north better than elsewhere," Berger writes in his report. "The native people there are a larger portion of the population than elsewhere

208

in Canada, and no provincial authority stands in the way of the Government of Canada's fulfilment of its constitutional obligation."

The fact that native people are—and will likely continue to be—the majority population in the Northwest Territories does provide an opportunity for a role in society which native people have not yet been able to attain anywhere else in North America.

But there are two possible approaches to this objective. The one advocated by the Dene leaders and Judge Berger calls for racially segregated societies in the north, with different political, economic, social and educational systems and institutions. The other approach is advocated by the NWT Legislative Assembly and by people like Peter Ernerk, an Inuit from Baker Lake, who is Minister of Economic Development and Tourism for the territorial government. It calls for a single, but pluralistic and multi-racial society, in which native people would have a much greater voice than in any other region of Canada. They would, in fact, likely wind up with the dominant voice.

Berger argues that a separate native society in the north, with its own political institutions, is required because efforts to assimilate native people into the mainstream of Canadian society have failed; because the territorial government is an institution that is foreign to native people; because the authority of the municipal-type councils which native people control in the small settlements does not give them the power to deal with their major concerns.

Berger argues that native people are different. "White people, in general, are driven by economic and social values that are very different from those that motivate native society," he writes. We must recognize "differences based on racial identity, cultural values and economic opportunities... The differences are real. They have always existed, but they have been suppressed."

Because of these differences, native people cannot be assimilated into Canadian white society. "Throughout Canada, we have assumed that the advance of western civilization would lead the native people to join the mainstream of Canadian life," Berger states. "Historical experience has clearly shown that this assumption is ill-founded... The statistics for unemployment, school dropouts, inadequate housing, prison inmates, infant mortality and violent death bespeak the failure of these programs."

Thus the need for a separate society. Northern native people "insist upon the right to determine their own future, to ensure their place, but not assimilation, in Canadian life," according to Berger. "Their concern begins with the land" and "extends to renewable and non-renewable resources, education, health and social services, public order and, overarching all these considerations, the future shape and composition of political institutions in the north."

"Only through transfer to them of real economic and political power can the native people of the north play a major role in determining the course of events in the homeland and avoid the demoralization that has overtaken so many Indian communities in the south."

The fact that nine of the fifteen members elected to the NWT Legislative Assembly are native people is dismissed, because "it is not regarded as a native institution," Berger states. "The native people see the Government of the Northwest Territories as a white institution ... For the most part, native employees hold clerical and janitorial positions."

Not everyone sees the differences the way Berger does. Harold Cardinal, at the time president of the Alberta Association of Indians, told the Berger Inquiry: "I think Indians and white people have similar goals in terms of what they want for their children. There is no family, Indian or white, that does not want to see a better home, a better lifestyle, and more success for their children than perhaps they as parents had during the course of their lifetime."

Colin Alexander, publisher of Yellowknife's *News of the North*, sees a danger in regarding native people as not only different, but as being all alike, rather than as separate individuals with separate thoughts and opinions:

There is a fundamental and tragic fallacy in thinking of the native people of northern Canada as a homogenous population with a uniform view of what the future should hold for themselves and their families ... as if they were some kind of special category of Canadians who should retain special privileges and special handicaps. There are two completely distinct aspirations of most of the native people of northern Canada ... Most individual native people want to be able to look other Canadians in the eye as equally self-sufficient and self-respecting individual citizens in confederation ... they want to achieve equal recognition among themselves and other Canadians of the view

210

that as a group they have a heritage as worthy of respect as do people of other backgrounds.

Dick Hill of Inuvik sees the challenge as establishing a society which can accommodate the widest possible range of life styles and aspirations. "There is a pluralistic society in most of the communities in the Northwest Territories," Hill told the Berger Inquiry. "Some of the residents are interested in hunting and trapping, others are interested in wage employment and business. Some are interested in both. Some want to continue living in traditional ways while some want to modernize. Some want regulations to protect their freedoms, while others want freedom from regulations. A flexible form of community government is required to accommodate the differences for the mutual benefit of the majority and the protection of the minority."

The NWT Legislative Assembly sees the allocation of governmental authority on racial grounds as tantamount to apartheid. "If you're for what Mr. Justice Berger seems to believe, then you've got to support something that has always been abhorrent to Canadians and violates our history," asserts a statement issued by the Assembly. "Frankly, support Mr. Berger and you have to support South Africa and its policy of apartheid—separate development for each of its founding races." William Thorsell, writing in the *Edmonton Journal*: "The word homeland. In South Africa, that is Prime Minister John Vorster's laundered term for apartheid—his policy of 'separate national development' for whites and each other 'founding nations' each with its own turf in the bosom of the South African state."

Professor W. H. McConnell of the University of Saskatchewan suggests that that which the Dene Nation leaders and Berger spurn is what native people in other parts of the world are struggling to attain. "While in parts of the world native peoples are unsuccessfully seeking membership in white-dominated legislatures, in the Northwest Territories" native people have already won a majority position in free elections, McConnell writes. Similarly the separate education that Berger advocates for native people in the north is the type of education that blacks in the United States struggled nearly a century to overthrow. McConnell recalls the historic 1954 ruling of U. S. Chief Justice Earl Warren which marked the end of U.S. school segregation, and Warren's finding that "separate . . . facilities are inherently unequal."

211

Peter Ernerk sees active participation by native people in local, territorial and federal governments as the best means to protect native cultural identity and political rights. "We want first and foremost to be full participating citizens of Canada, but we also want to exercise those rights and privileges which every Canadian citizen enjoys of active involvement in the political life of our communities," Ernerk told a 1977 conference on the Future of Canadian Federation in Toronto. He stressed, "our desire to preserve the heritage of our ancestors for ourselves, our children and their children. But we also want to contribute to the distinctive features of the Canadian mosaic, for we understand our role as people of native origin who are also citizens of Canada... We do not want to be 'sealed off' in separate political units built along racial lines. We do not need to be isolated from the rest of Canada in order to achieve our dual goals of political autonomy and preservation of cultural identity."

Everything that to me seems wrong with Judge Berger's approach was epitomized when, late in 1977, I met a young Indian school teacher in Inuvik. He taught both white and native students. What better way to learn, respect and benefit from the cultural values of different peoples? Under the Dene Nation concept, native teachers would teach native students, and white teachers would teach white students.

I was raised in a small community on the British Columbia coast, near an Indian reserve. The white students went to one school, and the Indian students to another. It seemed wrong to me then, and it seems wrong to me now.

Berger implies that white traditions, culture and values—other than "the selective adoption of items of western technology" to assist in hunting and trapping—have little to offer native peoples. But the converse of this is the most patronizing attitude of all: that native values have little to offer the rest of Canadian society.

Enerk was right in wanting to see native peoples "contribute to the distinctive features of the Canadian mosaic."

Never before in Canada's history have native people had the opportunity to participate as fully in the society and the mosaic in which they live, as they now have in the Northwest Territories. This will not be achieved in separate schools. Such a system can only lead to racial ignorance, racial intolerance and enemy camps.

Chapter 12
The Native Land Mystique

There is a mystique about the ideas of native peoples on the ownership of land. It not only colours the whole background of the Berger Inquiry, it has also to do with the future of Canada and the shape of our society.

The attachment of native peoples to the land is profound. Nothing is held more dear. The land is sacred. The land is owned by the Creator. People cannot individually own the land, they may only share in using it. The western social concept of individual ownership of land—of dividing, buying, selling and profiting from the ownership of land is a foreign idea, alien to the traditions, culture, philosophy and religious beliefs of native peoples.

This conceptual difference "is the heart of the matter throughout the whole world wherever native peoples find their land and their way of life in jeopardy as a result of intrusion and penetrations from the outside," according to Tony Belcourt of the Native Council of Canada. He was speaking for half a million native peoples in Canada when he appeared before Judge Berger in Ottawa, May 4, 1974, at one of a series of conferences to discuss procedures to be followed in the subsequent inquiry.

"All aboriginal people have a special relationship with the land that is religious in quality and which forms the foundation of their views of themselves as distinct peoples and nations," Glenn Bell, counsel for the NWT Indian Brotherhood, claimed in his final argument before Justice Berger in Yellowknife on November 19, 1976. Bell articulated the native position clearly: "only community ownership of the land, land which has belonged to our people for thousands of years, will give us the ability to determine and follow our own way."

The native concept of communal land ownership could contribute

213

to some radical changes in our society, and help achieve a more equitable distribution of wealth. But it is not a new idea, nor entirely alien to the western mind. In pre-feudal times, communal ownership of land was the norm, the last vestiges surviving in the commons of England and the communal pastures and farms of Europe. In the dim background of nearly three thousand years of western history is the drama of the conflict over whether land would be owned by the many or by the few; this is the conflict which Glenn Bell says is still today "the crux of this whole issue" facing northern native peoples. Social scientists, planners, economists, academics, reformers, are now taking a fresh look at the idea of public land ownership, not just in the context of native land claims but in the context of our entire society.

A startling thing about this idea of communal or public ownership of land and resources is the relationship it bears to the petroleum producing industry in western Canada. Perhaps nowhere in the world is there a more effective demonstration of how public ownership—at least of resources, if not land per se—can mesh with a free enterprise system and a relatively free market economy for the greatest public good.

But attached to the idea of this type of ownership is an unresolved question which may well prove central to the survival of Canada as a nation. Public ownership of land and resources by which public? By the Inuit people in the Arctic? By the Dene nation? By Alberta? Or by Canada? It is not a question which looms on the horizon. It is here, now. And as much as what happens in Quebec, it will be the answer to this question that determines what happens to the idea of Canada as a nation.

In Inuvik, Addy Tobac, a young university-educated Indian woman from Fort Good Hope and one of the leaders among her people, told me what the land means to her. "The blood and bones of my ancestors still live in the land," Addy said. "Their blood, their flesh, their bones: that is what the plants grow on, the animals feed on. My ancestors still live in this land. I am the land, and the land is me."

"We don't pretend to own the land in the same sense of private possession, giving us the right to exclusive use, to build fences, and put up 'no trespassing' signs," Tony Belcourt told Judge Berger at the pre-hearing conference in Ottawa. "In consequence, we don't feel that we have the right to do whatever we want with the land we occupy now and that our ancestors have occupied in the past. Your society might believe in that kind of system. We don't."

"We do believe, however, that as prior occupants, who inherited the land from our great-great-grandfathers, that we have the right to use it and pass on those rights to our great-great-grandchildren. We see ourselves as the guardians of the territory which we hold in trust for the generations that will follow."

George Manuel, former president of the National Indian Brotherhood, spelled out the native concept of land ownership in more detail in his book, *The Fourth World*.

In his travels, Manuel wrote, he found that throughout "the aboriginal world there has been a common attachment to the land. This is not the land that can be speculated, bought, sold, mortgaged, claimed by one state, surrendered or counter-claimed by another. Those are things that men do only on the land claimed by a king who rules by the grace of God, and through whose grace and favour men must make their fortunes on this earth."

"The land from which our culture springs is like the water and air, one and indivisible. The land is our Mother Earth. The animals who grew on that land are our spiritual brothers. We are part of that Creation that the Mother Earth brought forth. More complicated, more sophisticated than the other creatures, but no nearer to the Creator who infused us with life."

"The struggle of the past four centuries has been between these two ideas of land."

Native peoples speak with the passion and eloquence that comes from the depth of great anguish, and which makes the dispassionate rationalizations of educated white men seem dry and insipid. You don't breathe fire in writing a university thesis, nor preparing an argument in court. But Indians are now going to university, trading the eloquence of their ancestors for the cool erudition of the white professional. Judge Berger was treated to an example of this in a treatise on "A Concept of Native Title," presented before the inquiry by Leroy Little Bear of the Centre of American Indian Studies at the University of Lethbridge. After several centuries of white people trying to educate the Indians about the ways of civilization, Little Bear had turned the tables; in his words, he was writing to help establish "a better understanding of the Indians' property concepts" and "educating non-Indians."

Little Bear first reviewed the various means by which Europeans purported to claim title to Indian lands—by discovery ("simply setting foot on North America and planting a rag attached to a pole"), by adverse possession, by conquest, and by conveyance. He then' cogently argued that none of these means had been effectively employed to gain title to Indian lands in Canada. Next, he compared British and Indian property concepts:

An underlying premise of the British property system is that no one can own land in the same way that one can own a book. Since one cannot own land in the same way that he can own a book a system has been devised by the British to give symbolic ownership. This system is known as the estate system. Under the estate system one cannot outrightly own the land, mainly because land outlasts human beings. The land was there before the present owner, and will still be there after the present owner passes. Consequently, one can only have an interest in the land called an estate.

"A couple of observations can be made in regards to the estate system," Little Bear wrote. "Firstly, the system is linear vertically. The system is also very singular. It is geared to the individual ownership of land. Secondly, an underlying goal of the system is to facilitate transferability of the different interests. Thirdly, the system necessitates an extensive and complicated registry. It makes it possible to chronologically trace previous owners. If one went back far enough to the original source or original owner, one would discover that it is the Crown or Monarch." In other words, the source of title is the Crown.

"Indian ownership of property, and in this case land, is holistic. Land is communally owned.... If one attempts to trace the Indians' source of title, one will quickly find that the original source is the Creator. The Creator in granting land, did not give the land to human beings only but gave it to all living beings. This includes plants, sometimes rocks, and all animals. In other words, deer have the same type of estate or interest as any human being. This concept of sharing with fellow animals and plants is one that is quite alien to western society's concept of land. To western society, only human beings have a right to land, and everything else is for the convenience of human beings."

Little Bear concludes that Indian "philosophy, property concepts, and ramifications and implications thereof, may sound ridiculous and fairy-tale-like, but what philosophy does not? Do biblical stories make any more sense? To native people they sound rather ridiculous and make-believe. Does the Crown as a fictitious entity make more sense?"

Earlier, Judge Berger had explained to native peoples his concept of the Crown title system, in some brief comments at his hearings at Fort Rae. "The Government of Canada is sovereign over the whole of Canada," he said. "The Government is sovereign in the name of the Queen.... Under the constitutional law of our country, there is an underlying interest in all land that is held by the sovereign and that means in the name of the Queen. I have a house in Vancouver. It belongs to me but there is an underlying interest in the sovereign .... Under the constitutional law of every nation, the nation itself has an underlying interest in all of the land that comprises the country."

"When you come to think of it, the idea of land ownership does not make much sense," Manuel wrote in *The Fourth World*. "Anything else that a person might own is somehow or other the fruit of his labor. If the present owner did not make it himself, he exchanged what he did make for goods another person made... ownership of most things can always be related to the fruit of human effort. With land that is not so. Land and water are the sources of life. I might exchange my labor for the land that another man holds. But neither he nor any of the previous owners had a hand in the creation of that land. For him or me to claim more of a title to that than the right to use it while we have need of it is a presumption and an affront to our Creator as well as to our community."

Over the centuries a lot of other people have, as Manuel put it, "come to think of it." What many thought was that the mass of humanity was being oppressed by the owners of land. There are striking parallels in the chronicles of the land struggles of North American native peoples and Europeans of earlier times. There are similarities in the dispossession of people from land; the profound attachments to the land of the dispossessed; the religious and mystical connotations; the poignant eloquence that comes from the heart of suffering to catch the soul and stir the blood.

"The land shall not be sold for ever," is the Biblical injunction (Leviticus 25:23); "for the land is mine; for ye are strangers and sojourners with me."

The cities of ancient Greece sprang from a society where the land had been held by all, but eventually appropriated by the few. In Laconia, by 800 BC, the land was owned by an estimated thirty-two thousand citizens while another four hundred and fifty thousand toiled as slaves and tenant farmers. The concentration of wealth and grinding poverty erupted into revolution in which many of the rich were slaughtered and their lands expropriated. In Attica, wrote Aristotle, "A few proprietors owned all the soil, and the cultivators with their wives and children were liable to be sold as slaves on failure to pay their rent." Property was so valued, observed Plutarch, that "those that stole a cabbage or an apple were to suffer even as villains that committed sacrilege or murder."

The ancient Republic of Rome, too, rose from a community of free farmers. But by the time of Christ, most of the lands were owned by a few rich estate owners, the latifundium, while every year tens of thousands of slaves were brought from abroad. One of the first recorded attempts at radical land reform was led by the brothers Tiberius and Gaius Gracchus, who sought measures for the distribution of common lands to the poor peasants. The reforms were put down with swift violence by the patrician landlords, who offered equal weight of gold for the heads of these leaders. The head of the slain Gaius was filled with molten lead and brought to the Senate.

English peasants, revolting in 1381 against the fuedal lords who had usurped their lands, issued this demand: "We will that you free us for ever; us and our lands; and that we never be named and held as serfs."

In Oliver Cromwell's short-lived English republic, the mystic Gerrard Winstanley led a band of peasant followers who started to dig and plant the common lands with turnips and potatoes. "The earth shall be a Common Treasury of livelihood to the whole of mankind," Winstanley wrote in 1649. "This restraining of the earth from brethren by brethren is oppression and bondage; but the free enjoyment thereof is true freedom.... Let Israel go free, that the poor may labour the waste land and suck the breasts of their Mother earth, that they starve not."

Winstanley told Parliament: "The earth was not made purposely for you to be lords of it, and we to be your slaves, servants and beggars, but it was made to be a common livelihood to all ... the power of enclosing land and owning property was brought into creation by your ancestors by the sword; which first did murder their fellow creatures, men, and after plunder or steal away their land, and left this land successively to you, their children. And therefore, though you did not kill or thieve, yet you hold that cursed thing in your hand by the power of the sword; and so you justify the wicked deeds of your fathers."

Seven years later, in *The Commonwealth of Oceana*, Richard Harrington wrote: "If one man be sole landlord of a territory, his empire is absolute Monarch." But, "if the whole people be land-lords ... the empire ... is a commonwealth."

In the eyes of some defenders of capitalism, public ownership of land is communism. Abolishment of private ownership of land was, after all, called for in Karl Marx's *Communist Manifesto*. But if the father of communism was the enemy of private landowners, the father of capitalism wasn't much friendlier. Landowners grow rich in their sleep, Adam Smith pointed out in *The Wealth of Nations*: "That indolence, which is the natural effect of the ease and security of their situation, renders them too often ... ignorant." Rent is "the highest prices which the tenant can afford to pay." This leaves the landowner everything but "the smallest share with which the tenant can content himself ... and the landlord seldom means to leave him any more." Smith concluded that "every improvement in the circumstances of the society tends either directly or indirectly to raise the real rent of land, to increase the real wealth of the landlord."

Historically, communal ownership of land has been a common feature of all primitive societies, while individual land ownership has been a feature of more advanced societies. Nomadic people who lived by hunting and fishing had no need for individual ownership of particular pieces of land. But when men stopped to farm, establish trading centres, build cities, create industries, they needed, if not individual ownership of land, at least secure right to exclusive use of particular pieces of land in order to conduct these functions.

"The first man who, after fencing off a piece of land, took it upon himself to say, 'this land belongs to me,' and found people simple-minded enough to believe this, was the true founder of civil society," wrote Jean-Jacques Rousseau. In short, people had traded the advantages of communal ownership of land for the advantages of division of labour. How much of a net gain this trade-off provided for the vast majority of men may still be debated. "Man was born free," wrote Rousseau, "and everywhere he is in chains."

Why not capture the benefits of both concepts, the common ownership of land of aboriginal societies, and the greater production of wealth of the more advanced?

A method intended to accomplish exactly this was advanced just six years after Smith's *Wealth of Nations*, in 1782, by another Scottish professor, William Ogilvie of King's College, Aberdeen, in an obscure book, *Birthright in Land*. The essence of Ogilvie's idea was to allow the exclusive use of land necessary for enterprise and industry. In return for this privilege the unearned land rentals would be captured as a public tax. He argued that it is these revenues, "by which alone the expenses of State ought to be supported," so that "active, progressive industry should be exempted, if possible, from every public burden." Ogilvie's proposal anticipated by a full century the American economist Henry George whose single land tax system was one of the most widely espoused social reforms of the 19th century; a popularity that is echoed in Henry George societies that still exist today.

"Title to an equal share of property in land," Ogilvie argued, "is a birthright which every citizen still retains." This is denied by the monopoly of land, "a most oppressive privilege," worse than "all the tyranny of kings, the imposture of priests, and the chicane of lawyers taken together."

"When mention is made ... of the interest of any nation," wrote Ogilvie, "it is generally the interest of the landholders that is kept in view; nor would there be any mistake in this, if all men were admitted to claim ... their natural share of the soil."

Smith and Ogilvie wrote on the eve of one of the great dramas of a dispossessed people, the century long Highland Clearances of Scotland. It is ironic that the dispossessed of Scotland were in the vanguard of those who dispossessed Canadian Indians. In the saga of the Highland crofters and the native peoples are common tales of injustice, suffering and passionate eloquence.

Like the native peoples of North America, the early tribes of Scotland held land in common, and even when they became tenants of the clan chiefs, the crofters still regarded the lands that they used as theirs. They, too, had a profound attachment to the land, shrouded in religious mysticism. "Their affections ... are rooted, and grow like red buds of the coarse grass in the clefts of the rocks, out of their bare, bleak, wild mountain home," an observer is

220

quoted in *The Highland Clearances*, by John Prebble. "Their hearts are rooted to their hearths."

The clan chiefs found more money in sheep than destitute tenants, and the crofters were cleared out to make way for the "great improvements" of the lairds. The ministers of the church translated the eviction notices into Gaelic and read them to the people, soldiers burned the homes, and clansmen by the tens of thousands were left homeless, destitute, and starving. "Their was neither sin nor sorrow in the world for us," Prebble quotes one Peggy Mac-Cormack as writing. "But the clearances came upon us, destroying all, turning our gladness into bitterness, our blessings into blasphemy, and our Christianity into mockery." Betrayed by their chiefs, the Highland bards gave vent to the bitterness of the people:

And when a spade of turf is thrown upon you
the country will be clean again!
Or nothing will be placed over you
but the dung of cattle!

In the Southerland estates alone, some ten thousand clansmen were removed to make way for two hundred thousand sheep, while tens of thousands more were removed by other lairds. They scattered to the four corners of the world, but more than anywhere else they came to Canada. More than a century of evictions left twenty thousand Macdonells in Upper Canada's Glengarry County, and next to none but the laird in the Glengarry of Scotland. The flood of Scots emigrants swelled to a crest of nearly fourteen thousand in the single month of November, 1853. Packed like cattle in small sailing ships with stinking water and rotten food, thousands died before reaching landfall. For many, the hardships in the New World were as severe as those in the glens they had left. The first settlers arriving in the Selkirk colony (the Winnipeg of today) found themselves caught in the middle of a fur trade war, and once more the crofters were displaced, their homes burned again. If the arrival of the new-comers heralded the oppression of native peoples, it is difficult at times to distinguish between the oppressors and the oppressed.

Once a dominant theme of many proposed measures of social reform in the 18th and 19th centuries, the question of public land ownership is now generating renewed interest.

In November, 1975, while Judge Berger listened to the land

ownership concepts of native peoples, some three hundred civic planners and academics from across Canada and the United States were attending a public land ownership conference in Toronto, hosted by York University. (The conference proceedings are published in *Public Land Ownership: Framework for Evaluation*, Lexington Books, 1976.)

The opening address was given by Enrique Penalosa, Secretary General for the United Nations Conference on Human Settlement. One "consequence of the private ownership of land," Penalosa told the conference, "is the systematic impoverishment of the poor. In a time of rapid population growth and urban migration, the law of supply and demand applied to land produces an inevitable cost inflation. Rising land value is the greatest device for concentration of wealth in the world today. Uncontrolled pricing [of land] is therefore the greatest impediment to the more equal distribution of wealth, which is the pious promise of every society and its economic system."

"There is no justification for enrichment without effort. Since all profit must, in some way, come from effort, unearned profit must come from the efforts of others who are not protected from this exploitation."

According to Penalosa, "free enterprise could be stimulated or constrained regardless of whether the land is publicly owned or privately owned. It depends on how the land is managed, not on who owns it."

Ants Nuder of the Swedish Ministry of Housing and Physical Planning told the conference that before the 19th century, "private land ownership was unknown in Swedish towns and cities, where land belonged to the individual town or city or to the state." Municipal revenues were financed by annual ground rents, adjusted every three years to reflect changes in property values.

New York City, in 1645, set out a policy of selling only half the lands to be occupied, retaining the other half in a checkerboard pattern, the conference was told. The city-owned land was leased out on short terms which were renegotiated to reflect rising land values, providing the city with a growing source of income. For two hundred years the leased land provided so much revenue that there were no real estate taxes in New York City. Eventually, the city sold its land to finance construction of an aqueduct. The amount of revenue that New York would earn today if it still owned the land is staggering to contemplate. But the land was sold, and New York City is virtually bankrupt.

Another group that recently looked at the question of public land ownership was the Anglican Church's National Task Force on the Economy. Its 1977 report, "For the Elimination of Poverty and Social Injustice," followed four years of study and debate. It suggested that, "a large tax should be levied, in principle up to 100 per cent, on any increase in land values not directly attributable to the owner's investment in improvement."

"Even the most devoted upholders of free enterprise and private property have recognized, at least since the time of Ricardo (*Principles of Political Economy*, 1817), that income derived solely from land tenure serves no useful economic function and is nothing more than an unofficial tax levied by the landlord," the task force reported. It found that "there is no reason at all why the landlord should be allowed to pocket any part of the proceeds" from unearned land values.

When economists talk of land, they are speaking not only of the surface, but all the natural resources embraced by land: water, gravel, mineral ores, oil and natural gas. Seen in this light, the most valuable publicly-owned lands in Canada today are the oil and gas resources in the four western provinces. In the United States, most of the oil and gas rights are owned by the owners of the surface rights; in western Canada, the Crown, in the right of the provinces owns about three quarters of the oil and gas. The provincially-owned oil and gas resources are leased for private exploration, development and production. From their ownership of these resources the provincial governments of western Canada had, by way of production royalties and lease payments, collected net revenues of $17 billion by 1979, far more than the production profits of the oil companies.

Potentially, the oil and gas resources of the Northwest Territories and, to a much smaller extent, the Yukon, could rival and possibly exceed that of Alberta. An annual royalty revenue half as great as that collected by the Government of Alberta in 1978 would amount to some $30,000 per capita for the entire present population of the Northwest Territories and Yukon. That is fifteen times as great as the per capita oil and gas royalty revenues in Alberta.

As in the south, the oil and gas rights are owned by the public, and it is the public which will capture whatever economic rent is paid. But which public? Title now resides with the Government of Canada, or the Canadian public. But the Territories are pressing for eventual provincial status, and if they are accorded the same rights as the other provinces, it is they who would own the resources and reap the benefits. Native peoples of the north also claim ownership. The proposed Dene Nation wants full title to four hundred and fifty thousand square miles, including all the mineral rights. In the Arctic, the Inuit have proposed a settlement of their land claims under which they would share in the revenues, with a suggested three per cent royalty on oil and gas production. The Canadian Government has already indicated that in any land claims settlement native peoples would be offered a share in potential oil and gas revenues. But how this would be shared among different native peoples raises further questions.

Most of the potential oil and gas resources in the north, probably more than ninety per cent, are in the high Arctic, in the lands occupied by the Inuit. The Inuit people in the Mackenzie Delta region have asked for a three per cent royalty on any production from their area. Where would that leave the Indian and Metis people in the Mackenzie Valley, where the resources are very much smaller, but through whose lands pipelines would be required to permit production of the resources from the Mackenzie Delta region?

The point was not lost on Judge Berger. In Yellowknife, the judge wondered aloud whether in seeking a royalty the Inuit were "really saying that they are the only native people in the north entitled to share in the revenue of Mackenzie Delta gas because it happens to fall within the boundary of the territory that they say they have traditionally used and occupied? ... are not the Dene entitled as native people with a land claim based on traditional use and occupation to any share of Delta gas? ... Is the native share of that resource to go to the particular native group that happens to have used and occupied the particular area where the resource is found, or is the native share of the resources to be treated as something that ought to go to all native northerners on some per capita basis?"

The same question has to be asked in a wider context: Is the public sector share of the resource revenue to go just to the people of the north, or the people of Alberta, just because it happens to come from the area in which they live? Are not the people in Nova

224

Scotia and Quebec entitled to some share of that public revenue on some per capita basis?

Meanwhile, northern native people awaiting the settlement of their land claims can witness an object lesson of Adam Smith's dictum that "every improvement in the circumstances of society tends ... to increase the real wealth of landlords." Certainly the landowners of Whitehorse have already seen their real wealth increased from improvements anticipated from the planned construction of the Alaska Highway pipeline. "Land prices in Whitehorse," reported the *Financial Post* of December 17, 1977, "are rising to unprecedented heights, and most people believe it is just beginning ... rents have started to increase, some by as much as fifty per cent in the past several months. Industrial lots in and around Whitehorse have doubled and even tripled in the past four months [since the pipeline decision]. Light industry lots in the nearby subdivision of McRae were going for $3,000–$5,000 a year ago. Today, the same lot will fetch $17,000–$20,000. One prime fully-serviced light industrial lot closer to downtown is worth $90,000 on the open market. Four months ago, the government, which developed that area, had placed an upside price of $31,000."

The native concept of communal land ownership; and the echo of it that is faintly heard in the history of western society, listened to in the context of public revenues from petroleum resources, promises enormous consequences for the destiny of the Canadian nation. But how this is to be shared has yet to be determined. In the north, these revenues are still only potential, and the problem has not yet fully developed. Further south, however, these revenues are already flowing in the west at rates that could cause an economic hemorrhage in the rest of Canada. The final answers to these western and northern questions, just as much as what ultimately happens in Quebec, will determine the future of Canada.

Chapter 13
Why Alberta's Wealth Could Shatter Canada

Weep for Canada.

See it scattered like the pieces of a giant jig-saw puzzle.

The great brooding land, mother earth, the common treasury that Canadians thought was theirs is not. Now it's Quebec's land, it's Dene land. It's Alberta's oil and gas, not Canada's, Alberta's. Speak no more of the Canadian Rockies, for they are the Alberta Rockies.

When Albertans awoke to find that they were the wealthiest people in the country, their spokesmen and leaders became paranoid. The greater their wealth and power became, the more determined they seemed to seek retribution for past wrongs, both real and imagined, that they had suffered at the hands of eastern Canada.

The westward progression of industry, trade, people and wealth is inevitable, historic, a fleshing out of the Canadian nation. The decision to approve the Alaska Highway pipeline was seen as a quantum jump in the shift of power to the west. But the industrialization of the west and wealth this produces are not the problem.

Rather, the problem flows from the tens of billions of dollars in unearned tribute that Alberta bleeds from the rest of Canada. It is an economic hemorrhage which will drain the country. If not staunched, it will terminate the already fractured alliance of self-seeking regions that now constitutes our nation.

From its ownership of most of the province's oil and gas resources, the Government of Alberta will collect some $4 billion in net revenues during 1979, an amount equal to $2,000 for every Albertan. During the 1980s this revenue could exceed $25,000 for every Albertan, $75,000 per family. By 1990 the Government of Alberta stands to have a pool of investment capital as great as all American investment in all of Canada in 1977. Before it finally backed off, the federal government tried in 1974 to capture some of Alberta's oil wealth for the benefit of other Canadians, a move that nearly created a crisis. Another confrontation, this time perhaps the showdown, is inevitable—soon. If this fails to result in more equitable distribution of oil and gas wealth, the idea of a Canadian nation seems doomed.

Only a nation-wide, equitable distribution of this type of wealth, whether it comes from Ontario, the Arctic, or Alberta, will maintain the kindred spirit without which Canada cannot endure.

When the showdown with Alberta comes, Ottawa will be missing a major card—a Mackenzie Valley pipeline to tap federal energy resources in the north. The pipeline decision prevented the development of the largest, most economically feasible block of energy outside of Alberta, a resource that otherwise could have been in production by 1977. The billions of dollars that would have been invested in the north are now all focused in Alberta. The pipeline decision strengthened Alberta's already powerful hand, and Alberta did not even seek this extra ace.

In Edmonton, Alberta Energy Minister Don Getty told me he didn't think the gas pipeline decision affected the province's position "one way or the other." But, he said, "I do think it's bad for Canada. I think the Berger Commission recommendation is a mistake. It doesn't just shut down gas, it shuts down oil, too. I don't see how you could build an oil pipeline if you can't build a gas pipeline."

Frustrating the development of federal resources certainly does nothing to quell the clamour of those Albertans who are demanding a new deal in Confederation. Often, the loudest calls seem to come from those who have newly arrived in the western promised land—there is no zeal like that of the convert.

Here, for example, is Edmonton MP Douglas Roche, a former eastern journalist who moved to Alberta in 1955, writing in *Weekend Magazine* (December 3, 1977):

> Albertans know that having energy in an energy-hungry world has finally given us the power to demand, not plead, for a renegotiated confederation to ensure our self-reliance. Power. That's what the Alberta story is all about.

Here is Bob Blair, speaking to the Canadian Club of Hamilton, in November, 1977:

Sometimes when hearing some more doom and gloom conversations some of us have said in Ontario, 'However we do happen to have lots going on,' and we got the immediate answer, 'Yes, sure, but that is in Alberta,' as though that kind of refuted ... our proposition that everything is not all bad. That is still a lingering signal of the regional bias that exists in this country; that good results are only significant if the manna drops evenly, between Sarnia and Montreal.

That provokes one point which deserves to be hit every time. Even if the action were confined within Alberta, it is still darn well in Canada ... If someone should have to travel across to the west to make a sale or get a job that is hardly fatal to our nation. A bit of regional shift is not catastrophic. There have been many generations of prairie kids coming down to the Sarnia-Montreal zone to find an industrial or head office job, and plenty of corresponding purchasing money moving in the eastward direction. So perhaps some of the kids, and the capital, may drift back where they were generated. It may be surprising, but it is not un-Canadian. It is all still very much in Canada.

Is it really very much in Canada? Is it really that much better for Quebec than oil from Saudi Arabia or Russia if Alberta is going to charge the world price (as it should) and at the same time pocket (as it shouldn't) all of the economic rent for itself?

Many Canadians fear that even if Quebec does not opt out, Canada is becoming so decentralized that it will cease to function as a nation.

Quebec Finance Minister Jacques Parizeau claims Canada is already the most decentralized of all federations and doubts its survival if further decentralization occurs, " ... because rather often in Canada we tend to talk of the abusive centralized powers of Ottawa, we tend to forget that in reality Canada is highly decentralized," he states (Toronto *Globe and Mail*, December 9, 1977). "And yet some people talk of more decentralization to appease Quebec. Can we afford more decentralization without rupturing the country? When does a central government cease to be a central government?"

Syndicated columnist Richard Gwyn claims that, "Alberta Premier Peter Lougheed wants a decentralized Confederation that would differ little in substance, if not in symbolism, from Quebec Premier Rene Levesque's version." Surveying the results of the three-day conference of First Ministers in Ottawa in February, 1978, Gwyn concluded that, "Canada is less and less one country, or even two nations, than a Confederation of 10 something or others." He predicted that by the time the First Ministers meet again the "transformation of Canada from a country to a Confederation—an association of semi-sovereign states in fact—will be made explicit."

One of the significant decentralizing forces in Canadian federation was the granting of Crown lands, including mineral rights, to the provinces when the country was founded in 1867.

The British North America Act states: "All lands, mines, minerals and royalties belonging to the several provinces of Canada, Nova Scotia, and New Brunswick at the Union, and all Sums then due or payable for such lands, mines, minerals or royalties shall belong to the several provinces of Ontario, Quebec, Nova Scotia and New Brunswick in which the same are situated or arise."

According to Gerard V. La Forest (*Natural Resources and Public Property Under the Canadian Constitution*), the Fathers of Confederation expected that revenues from these lands would provide the provinces with "adequate revenue for performing their limited functions." However, it was different in the prairie provinces where the federal government retained title to all Crown mineral rights, including oil and gas resources, until 1930. When Canada purchased what are now essentially the three prairie provinces from the Hudson's Bay Company in 1870, it also acquired the mineral rights. In La Forest's words, it was "clearly envisaged that the resources of the area would pass to Canada to be administered for the benefit of the whole country." The BNA Act of 1867 anticipated the future incorporation of the province of Manitoba (which took place in 1870) and provided that, "All ungranted or waste land in the province shall be, from and after the date of the said transfer, vested in the Crown and administered by the Government of Canada for the purposes of the Dominion." The two acts creating the provinces of Saskatchewan and Alberta in 1905 provided that, "All Crown lands, mines and minerals and royalties incident thereto ... shall continue to be vested in the Crown and administered by the Government of Canada for the purposes of Canada."

The fact that they were not allowed to own the mineral resources like the other provinces contributed in no small measure to western feelings of alienation and exploitation. After much agitation, the prairie provinces finally acquired these rights in 1930 by an amendment to the BNA Act which states: "In order that the provinces [Alberta, Saskatchewan and Manitoba] may be in the same potition as the original provinces of Confederation . . . the interest of the Crown in all Crown lands, mines, minerals (precious and base) and royalties derived therefrom within the Province . . . shall . . . belong to the province." The west achieved some equity, but national unity would have been better served if that equity had been achieved not by granting mineral rights to the prairie provinces, but by granting the mineral rights of all the provinces to Ottawa. Perhaps if they had any inkling of how much wealth western mineral rights would produce, the other provinces would have found this acceptable.

It has always been in Alberta that western alienation and resentment against real and imagined discrimination by the power interests of the east have been felt most strongly. For years Alberta was a poor boy of Confederation. Albertans resented the high prices they had to pay for the benefit of tariff-protected manufacturers in the east. Because of the longer rail shipping distances, Alberta paid even more than Saskatchewan and Manitoba for goods brought from the east, and received less for agricultural products shipped to the east. In the thirties, Alberta was hit hard by everything at once. Plagues of grasshoppers blackened prairie skies, were ground to grease on the railway tracks, and stopped the trains. The drought was so bad that locomotives in southern Alberta at times used snow-ploughs to push through drifting topsoil as deep as ten feet. The Depression tumbled the price of beef in the Calgary stockyards from seven to two cents a pound. By 1935, Alberta farmers had accumulated debts (mostly to banks controlled in the east) amounting to more than three hundred million dollars; the interest alone amounted to one-third of the total farm income. Even by 1945, personal income per capita in Alberta was still nine per cent below the national average.

It is different today. Alberta is the only province in Canada with no debts, no inheritance taxes and no provincial sales tax. It has the lowest energy prices, the lowest personal and corporate income taxes, the lowest property taxes, and the highest per capita expenditures for education, health and social services. Alberta's gross provincial product had risen, by 1976, to eleven thousand one hundred dollars per capita versus nine thousand three hundred and fifty in Ontario, and this gap will widen rapidly.

"There is a shift in Canada to the west," says Lougheed. "The population is shifting out west... our indigenous market is growing pretty rapidly. We're going to get to the point in this country where we really have so much economic viability going on in western Canada that the local market is going to start playing a larger and larger role."

For all that there is still a touch of paranoia in continued complaints about discrimination in national economic policies and constant apprehension about the menace of federal government intrusion. This has had a bearing on the western attitudes to the pipeline proposals.

I interviewed Lougheed early in 1971, shortly before he led the Tories into power in Alberta, and while I was still editor of *Oilweek* magazine. Even then, Lougheed expressed concern about the possibility of "federal government manipulation" to divert petroleum industry expenditures away from Alberta to federal lands in the north. And he hinted that Alberta would seek to control the flow of any federal gas from the north flowing by pipeline across the province. "It has to be in Alberta's interest to have the gas flowing through Alberta with Alberta a key part of the transportation system," he said. "The ultimate approved project should utilize the existing Alberta transmission facilities, from Alberta's point of view."

Two years later he told a CBC radio audience: "We have to be concerned because northern gas and oil, of course, is within federal jurisdiction and we have to protect the interest of Albertans and assure that any national energy plan is not established in such a way as to put Alberta development into a secondary position."

When the pipeline applications were before the National Energy Board, Alberta adopted a position of neutrality, the only province west of New Brunswick that did not intervene in the hearings before the board. Lougheed explained this position in a statement to the Alberta Legislature in June, 1975: "We don't think it would be appropriate for the Alberta Government to take a specific position with regard to the two competing applications at this time, although that position is subject to reconsideration... We are, however, inclined by preference toward the Maple Leaf project, because we understand it would more extensively use the existing facilities of transmission and existing Alberta facilities generally, and I believe too, would reflect support for a company under charter by the law of the Province of Alberta."

After the pipeline decision, an editor with one of the largest western daily newspapers told me: "Most westerners hope that Canada will stick together. But if it doesn't, we want to make sure that we can look after ourselves. That's why giving control of the pipeline to the western companies is considered a victory for the west."

"Albertans do have a story to tell," writes Keith Spicer, formerly Canada's Official Languages Commissioner, in the *Toronto Star*, November 30, 1977. "They may 'lose' $500 a year per capita because they're a captive market for high-cost eastern manufacturing. They pay about $300 a year per capita in federal taxes to bail out have-not provinces. Perhpas stretching a point, they claim they are even sacrificing $4.5 billion over 10 years to give other Canadians cheaper gasoline."

By agreeing to phase in over a period of years the rise in Alberta oil and gas prices to world levels, Lougheed says Alberta is providing a subsidy to other Canadian consumers. "The history of confederation," he says, "is going to show, I think, that in this period of time no province has made a greater contribution to confederation in economic terms, by the billions."

In his *Weekend* article, Edmonton MP Douglas Roche writes: "You can't live in Alberta very long before realizing that the wave of affluence attracting the attention of the rest of the country conceals two basic facts: The terrible suffering of the drought and Depression years, and national economic policies that continue to discriminate against the west and make Albertans fearful that once the oil runs out they'll have nothing left to bargain with. It takes a long time to forget the 1930s; selling wheat at 23 cents a bushel or shipping a 900-pound steer and getting a bill for freight in return."

It may take a long time to forget the 1930s, but it hasn't taken long to forget the 1960s. That was when the world was awash with surplus supplies of oil priced so low that, as Prime Minister Trudeau recalled in 1973, "a barrel of petroleum from Saudi Arabia could be delivered in Edmonton at a lower price than a barrel of oil produced in an Alberta well just 40 miles away." To prevent the collapse of the Alberta oil industry, the federal government established the National Oil Policy, which kept the lower-priced imported oil out of Canadian markets west of Ottawa. The result was that in the 1960s, consumers in Ontario and elsewhere subsidized Alberta to the tune of at least $500 million dollars in the form of higher energy prices. Not all national policies discriminate against Alberta.

Perhaps one reason that Lougheed appears forever vigilant against the menace of federal intrusion is that the control of its oil and gas resources, which Alberta so fully exercises, may rest on shaky constitutional grounds.

It is more than ironic that the first prominent questioning of Alberta's right to regulate the production and sale of its oil and gas came from one John Ballem, Calgary lawyer, author, counsel for Imperial Oil, Gulf and Shell at the Berger and National Energy Board hearings, and a former law partner with Peter Lougheed. In an article, "Constitutional Validity of Provincial Oil and Gas Legislation," in the *Canadian Bar Review* (1963), Ballem examined a lengthy string of legal precedents which suggest that the power exercised by Alberta belongs to Ottawa. Ballem pointed out that the BNA Act provides the Parliament of Canada with exclusive legislative authority over:

- "the regulation of trade and commerce" which extend beyond the bounds of a single province; and
- connecting works ("lines of steam or other ships, railways, canals, telegraphs and other works and undertakings connecting the Provinces with any other of the Provinces or extending beyond the limits of the Provinces.")

In addition, Ottawa has the power to assume control by declaration (control of "Such works as, although wholly situate within the Province, are before or after their Execution declared by the Parliament of Canada to be for the General Advantage of Canada or for the Advantage of two or more of the Provinces.")

Ballem also noted that where provincial and federal legislation overlap under BNA Act terms, "the Dominion Legislation must prevail."

Ottawa's declaratory powers can be applied with stunning ease, and have been nearly five hundred times. Ballem noted that when the Supreme Court found in 1925 that the federal government could not regulate grain elevators within a province, which were subject to provincial legislation, Ottawa simply issued "a remarkably straightforward" declaration stating: "All elevators in Canada heretofore or hereafter constructed are hereby declared to be works for the General Advantage of Canada."

Alberta's position has been that its rights as owners of oil and gas resources transcend Ottawa's rights under these BNA Act provisions. But Ballem claimed that Alberta's control "as an incident of ownership... may still be open to attack on constitutional grounds."

"The declaratory power, used in conjunction with trade and commerce, and interconnecting works, could enable the Dominion to exert virtually complete legislative control over the industry," Ballem concluded.

Lougheed, of course, has a different view on these matters. "The constitutional implications are that the ownership of oil and gas rests with the provinces," he has said. Period. End of debate.

Just to be certain, however, Lougheed's demands for "our valid claims for a new deal in Confederation" include, preeminently, the entrenchment of provincial rights over resources in any patriation of the constitution.

No one has ever taken Alberta to court to test the validity of its oil and gas legislation, but ever since the Ballem article the possibility of a constitutional crisis involving Alberta's energy supplies has loomed in the background. "We almost had it in 74, with the oil," Lougheed has said. "Yes, it could happen,"

What almost happened in 1974 actually started earlier in a running battle between Alberta, Ottawa, and Ontario over Alberta's oil and gas supplies, prices, revenue sharing and plans for world-scale petrochemical plants.

The war's first shot was fired by Alberta in 1972 when it set up a two-price system for its natural gas, with higher prices for Ontario and other consumers, and lower prices for Albertans. Well, not exactly a two-price system, because that would have been too clearly a discrimination in inter-provincial trade and an invasion of the federal turf. Instead of lowering the prices paid by Alberta consumers, the provincial government provided rebates on the gas bills. You see, it wasn't really a two-price system, after all. But Ontario's Provincial Secretary for Resource Development, A.B.R. Lawrence, was not soothed. He called Alberta's gas price policy "so serious that it shakes the foundation of confederation itself."

Things got a lot hotter after the Middle East War in 1973, the Arab oil embargo, and the sudden quadrupling of world oil prices, all of which pulled Alberta's prices up in tandem.

In September that year, Ottawa slapped a tax of forty cents a barrel on Alberta oil exported to the United States. The tax later rose to more than five dollars, but it is now coming off as Alberta oil prices gradually climb to world levels. Lougheed called the Ottawa export tax "the most discriminatory action by the federal government against a particular province in the history of confederation." He said it was a "power play to control Alberta's oil and gas resources; we intend to fight back. We are going to have some pretty strong responses."

As its oil prices climbed, Alberta moved to shut off a potential windfall for the oil companies and capture increased revenues for itself. It slapped royalties on the increased prices up to sixty-five per cent. But Ottawa wanted a piece of the action, and Finance Minister John Turner intended to carve out the federal share in provisions of the federal budget introduced in the House of Commons on May 5, 1974. The budget would not have allowed oil companies to claim royalty payments to the provinces as an expense in computing federal income taxes. The Canadian Petroleum Association estimated that the combined effects of the Alberta and Ottawa actions would be to capture government revenues of $9.13 per barrel from oil sold in the export market for $10.50 per barrel. Imperial Oil president W. O. Twaits said the governments were like "two hogs under a blanket, fighting for the same acorn." Caught in this squeeze, the oil companies (not just the multinationals but the Canadian-owned companies, too) started pulling their drilling rigs out of Canada, and exploratory drilling dropped twenty-five per cent in 1974, at a time when increased prices should have stimulated exploration investment.

In a later address to the Calgary Chamber of Commerce, Turner defended his budget proposals, claiming "we seek only a modest share of oil revenues. It is the provinces which are claiming the lion's share." Turner vowed that "the federal government would not be squeezed out of access to taxation of the oil industry ... the federal government is the only government in Canada that has the power to re-distribute wealth between rich and poor regions ... To ensure that it continues to have the fiscal resources to exercise its national responsibilities, no Canadian Parliament can allow its access to any major source of revenue to be usurped by provincial action."

Two days after Turner introduced his budget, it was defeated in the House by the Conservatives and New Democrats, and the country went to the polls. A detente, of sorts, seems to have been reached by that November when a new federal budget was introduced. It again prohibited oil companies from deducting royalty payments, but Ottawa reduced its tax bite so that the federal victory was more one of principle than profit. Alberta, too, eased off its high royalty rates and agreed to reimburse most of any extra expenses that did fall to the oil companies because of Ottawa's non-deductibility provisions. The oil rigs came back, and exploratory drilling increased one hundred and twenty per cent in three years, resulting in encouraging new discoveries.

The war of words, however, had not quite subsided. As late as May, 1976, Ontario Premier Bill Davis suggested in the Ontario Legislature that disputes over energy prices could be quickly resolved if the federal government were to invoke BNA Act provisions to declare Alberta's oil and gas wells to be "works for the General Advantage of Canada." Two days later, Lougheed responded in the Alberta Legislature: "We give notice to the Ontario Legislature that my colleagues and myself, and I believe the people of Alberta, simply will not accept a federal government attempting to expropriate resources owned by Albertans for the benefit of central Canada." He declared that in his view such action would mean that Canada would no longer be a federal nation, nor a nation that Alberta would want to be a part of.

That November, the victory of Rénè Lévesque's Parti Québecois at least temporarily shifted the Confederation crisis away from a confrontation with Alberta. Edmonton and Ottawa suddenly emerged as allies for a number of reasons. Girding for the fight with Quebec and dealing with the problems of an ailing economy, the Trudeau government needed the co-operation of the provinces. This was no time to do battle with Alberta. Besides, Alberta's quest to increase its oil and gas prices to world levels accorded with the federal government's strategy to avert serious energy shortages. The higher world oil prices were seen as key to stimulating supply, curbing demand growth, and fostering conservation. For his part, Lougheed undoubtedly welcomed any respite in the incursions from Ottawa. In addition, he needed Ottawa's help to negotiate lower tariffs for Alberta's agricultural and petrochemical exports to the United States.

But Lougheed paid a price for his alliance with Ottawa, and it rankled. That price was the establishment in Sarnia, Ontario, of a

236

world-scale petrochemical complex by a group of companies headed by the federal government's Polysar Corporation which would use Alberta's oil. Lougheed had long had his sights set on world-scale petrochemical facilities in Alberta, and the Petrosar complex would capture much of the domestic market, slowing Alberta's plans. Because it is much cheaper to transport crude oil to the major market centres than it is to transport the finished petrochemical goods, economics dictated that the complex should, indeed, have been located at Sarnia. But bacause it was Alberta's oil, Alberta politics dictated that it should have been in Alberta.

"There won't be another Petrosar," Lougheed vowed in late 1977. "There will not be another development using Alberta crude oil or Alberta natural gas located in Sarnia. Petrosar is the last one ... If we ever had another Petrosar I really think the reaction in Alberta would be explosive. That's literally taking jobs away from the people in the west and shipping them down the pipeline ... We're really upset about that ... That's just unfair in the extreme. It makes the development of our petrochemical industrial complex much more difficult. And it's damn disturbing."

It is also damn disturbing to think about the implications of what Lougheed is suggesting. Implicit in his idea are restrictions in the freedom of capital and people to locate wherever in Canada is most advantageous. The very essence of a nation demands complete freedom in the internal movement of trade, investment and people.

But if Alberta lost out to Petrosar, it got sweet revenge with the decision to award the Alaska Highway pipeline to Calgary's AGTL-Foothills organization. This is seen as the prime lever which will shift the power out of the hands of the resented easterners and into the hands of those eager Albertans. Everyone says so.

Vancouver's Allan Fotheringham, writing in *Maclean's*, has said, "...half of Toronto is going to collapse in a psychic faint when it realizes what the decision to give the Alaska Highway pipeline route to the Foothills group of Calgary is going to mean to the traditional power base of Canada."

Edmonton's Douglas Roche: "Alberta is remaking the economic map of Canada...A northern gas pipeline is coming which will consolidate Alberta's position as the financial centre of western Canada."

The Toronto *Globe and Mail*, noting that AGTL is at the centre of both Alberta's petrochemical plans and the northern pipeline:

"AGTL is the key company in the business activity that is swinging the economic power in the country out of eastern Canada and into the west."

Bob Blair, quoted in *Maclean's*: "The decision has wrenched power away from some people who have become accustomed to having power. The big oil companies, for example, and certain parts of the Toronto financial establishment."

The fact, however, is that the pipeline decision is almost incidental to the real cause of the shift to Alberta. Whatever wealth, or power might accrue to the west because lead sponsorship of the pipeline venture may reside in western companies, it will be peanuts compared to the economic rent captured by Alberta from its energy resources.

To understand how "economic rent" accrues from Alberta's oil and gas, sweep away the confusion that now surrounds the use of this term. The Alberta situation is a classic example of economic rent as first defined by David Ricardo one hundred and sixty years ago. Ricardo understood that some lands produce more than others. If it cost two dollars a bushel to produce wheat from the worst land being farmed, but one dollar from the best land being farmed, then the best land will earn the difference, an extra profit, or, in short, economic rent. As long as the demand for wheat could be satisfied from the best land, the price for wheat would be only one dollar and there would be no economic rent for anyone. But when it also becomes necessary to farm the worst land in order to have enough wheat, then the price of wheat will rise, and the owner of the best land will capture the economic rent without any labour or effort.

This is just what has happened with oil and gas. There are still a lot of conventional, low cost oil and gas resources, but not enough — in Alberta or anywhere else in the world — to meet the total demand. So energy prices bounce up to meet the cost of new and far more expensive supplies, such as the Athabasca oil sands, heavy oil deposits and others. And the difference in the cost of these forms of energy yields enormous economic rent for the owner of the lower cost reserves, which in Canada principally means the Government of Alberta.

The four western governments collect their resource economic rent through production royalties, lease rentals and cash bids at lease auctions. In the thirty years to 1978, this economic rent (which is in addition to all forms of taxes) has totalled $13.7 billion. Alberta's share was $11.5.

The worry that Alberta's oil revenues might run out in a few years and that it needed to collect a nest egg against this rainy day made all this largesse seem acceptable. Knowing it would not last, few begrudged Alberta's efforts to prepare for the next drought.

Now it is starting to dawn that Alberta's oil and gas revenues will not run out in the foreseeable future; in fact they will increase substantially. Declining production of conventional crude oil has been temporarily reversed as a result of recent discoveries. When this decline is resumed, it will be offset by higher prices, increased production from the Athabasca oil sands, and increased natural gas production throughout at least the 1980s.

Every gallon of gasoline purchased by a motorist in Ontario (or in most other parts of Canada) means a net profit of some twentyy cents for the Government of Alberta. And Lougheed means to keep it all.

Chapter 14
The Alaska Highway Pipeline—Dead or Deferred

As the U.S. Congress in September, 1977, pondered whether or not to ratify the decision of President Carter to approve construction of the Alaska Highway pipeline, the *Wall Street Journal* registered a timely note of caution.

"The Senate needs to understand that the much-touted pipeline to bring gas from Alaska will never be built," said the lead editorial in the *Wall Street Journal*. It added: "Oh, never may be a mite too strong."

Events in the following months proved this caution to be well-founded.

Before the middle of 1978, it had become clear that the proposed Alaska Highway line was either dead, or deferred for at least five to ten years. No other outcome seemed conceivable. The amount of money required was clearly more than U.S. and Canadian pipelines could borrow by themselves if all of them decided to participate in the project. Most of them, in fact, decided not to participate. The gas supply situation had dramatically changed, from one of crippling U.S. shortages and a threat of similar Canadian shortages, to surplus supplies, although the U.S. surplus may be temporary. Large Mexican gas reserves had emerged to offer the United States an alternative supply at a much lower cost than Alaskan gas. Finally, deferring construction of the Alaska Highway pipeline offered the prospect of delivering much greater volumes of Alaskan gas at a later date, thereby reducing the transportation cost.

Doubts about the ability to economically move North Slope gas along the Alaska Highway were sounded before the pipeline was approved in principle in September, 1977, as the *Wall Street Journal* editorial suggests. This doubt was first raised as a question about whether or not the pipeline could be financed. C.O. Goldsmith, vice-president of finance for Atlantic Richfield Company, testified at hearings before a U.S. Congressional committee on October 14, 1977, that there was "a substantial likelihood of delay because the sponsors may be unable to raise the enormous sums required under a financing plan which currently lacks sufficient credit support." Goldsmith said he had "talked to the senior lending officers, the senior executives and vice-presidents of the largest commercial banks in the United States. I have talked to other investment banking firms... In all cases all of these people are extremely sceptical as to the ability of the project to obtain the quantity of capital required and sceptical about making loans to it."

Representative W. Henson Moore of Texas, in a supplemental view attached to the House Committee's report recommending approval of the President's decision, said he was "very sceptical that this pipeline can be constructed with private financing." He predicted that "the corporations constructing the pipeline will surely turn to Congress seeking Federal grants, loans or loan guarantees."

Congressman Carlos Moorehead of California expressed surprise that the giants of the gas business seemed to be having difficulty raising a few paltry billion dollars. "I think that those of us who are responsible for spending $460 billion to $480 billion in a year have some difficulty understanding why you guys are having so much trouble with fifteen billion bucks," he commented during committee hearings. "I guess the difference is, we have the printing press."

Behind these, and other similar comments, usually lay a common assumption. If only someone—the State of Alaska, the oil companies, the Canadian or U.S. governments, even the Government of Alberta—would assist in borrowing the money, the pipeline would be a going concern, selling great quantities of gas to buyers desperately in need of energy. But what if no one wants to buy the gas for what it will cost? What if alternative energy supplies are available at a lower price?

Any government which stood behind debts of up to $15 billion and then discovered that the undertaking could not pay its way, would wind up with the biggest financial mess of all time. And yet it now seems clear that the United States really does not need Alaskan gas at the price it would cost delivered by an Alaskan Highway pipeline before the late 1980s or early 1990s.

It was exactly this view that was raised by six Congressmen in a minority view attached to the House Committee report on the

241

Alaska Natural Gas Transportation System, in October, 1977. "This 4,787-mile project will be the ninth wonder of the world—if you consider the Alyeska oil pipeline as the eighth—and it will cost accordingly," they stated. "The prospective builders and operators of this pipeline have already indicated their inability to finance this amount without help from some other source of revenue than their own." They concluded: "One also has to wonder whether this costly Alaskan pipeline might not have to be built now if the price of all domestic natural gas were deregulated. If such deregulation encouraged production of supplies adequate to fill all existing pipelines in the lower 48, it might be many years before the question of financing the ninth wonder of the world would have to answered."

Robert Herring, chairman of Houston Natural Gas Corporation, predicts that there will not be an adequate market for Alaskan gas before 1987 or 1988, four to five years after the originally scheduled completion of the pipeline. At that, he might be optimistic. In an address to the Independent Petroleum Association of Canada at Calgary in March, 1978, Herring said that the unregulated U.S. gas supplies at prices of $2.20 per thousand cubic feet are losing markets to lower-priced coal supplies. A recent switch from natural gas to coal at just one thermal-electric power plant in Louisiana represented a loss in gas sales amounting to $700,000 a day, according to Herring. He estimated that Alaskan gas will cost $5 per thousand cubic feet, more than double the 1978 price for unregulated U.S. gas supplies.

Indicative of the difficulties confronting the Alaska Highway pipeline is the problem that Trunk Line, Westcoast Transmission and Northwest Pipeline have had in attracting other firms to join them as sponsors of the project. In April, 1978, TransCanada Pipelines announced that, following nine months of discussions, it had decided not to join as a sponsor. None of the seven Canadian firms participating in the Arctic Gas project at the time of the National Energy Board decision, and only three of the eight U.S. firms, had joined as sponsors of the proposed pipeline. In total, eight firms, including Trunk Line, Westcoast, Northwest Pipeline and five other U.S. pipeline firms, were sponsors of the Alaska Highway project in early 1979. Not since the initial studies in 1969 had there been so few firms involved in the planning to transport Alaska North Slope gas.

When approved in 1977, the Alaska Highway pipeline was estimated to cost $10 billion, and gas deliveries were to start by January 1, 1983. In November, 1978, Mitchell Sharp, former federal cabinet minister and head of the federal government's Northern Pipeline Agency, told the House of Commons Natural Resources Committee that completion would be delayed until the Fall of 1983, and that the estimated cost had increased to $12 billion. In February, 1979, the project sponsors announced that the target completion date had been further delayed to November 1984. Estimated cost was reported to have increased to $14 billion. In addition to the pipeline itself, a further investment of some $2 billion will be required for a processing plant to treat the gas at Prudhoe Bay, as well as substantial investments to produce the gas. Total investment for production facilities, the processing plant and the pipeline are conservatively estimated in a range of $15 billion to $20 billion. That compares with a total investment of $12 billion to place Prudhoe Bay oil on stream in 1977, providing a daily energy supply four times as great as the projected production of Prudhoe Bay natural gas.

By early 1979, the only firm supply assurance that existed for the proposed pipeline was a volume of two billion cubic feet per day from Prudhoe Bay, less than half the design capacity of the pipeline. Operating a pipeline at less than half of the design capacity does not provide low-cost transportation. The annual cost—including earnings, interest, fuel and operating costs—is certain to exceed twenty per cent of the investment, or $8 million to $12 million a day. With a supply volume of only two billion cubic feet per day, a delivered cost of $4 to $6 per thousand cubic feet seems a conservative estimate.

By contrast, the 1978 U.S. market price for natural gas was only $2.20, and about the same for imported oil. Liquefied natural gas imported by tankers was available for about $3.50 and potential suppliers in the Middle East, Africa, Indonesia and elsewhere were glutted with surplus gas for which they had no markets. Late in 1977, the U.S. government rejected a planned forty-eight inch pipeline intended to supply the United States with large volumes of natural gas from Mexico, because the offered price of $2.60 was considered too high. In early 1979, U.S. President Jimmy Carter and Mexican President Jose Lopez Portillo renewed negotiations on the sale of Mexican gas.

The United States, however, can use Alaskan gas if the cost is competitive with the cost of other energy. By 1979 the United States was paying more than $100 million a day for a rising tide of imported energy, mostly crude oil but also natural gas imported from Canada and, in liquefied form, from other countries.

A large capacity pipeline with enough supply to operate at full rather than half capacity is probably the only economical way to transport Alaskan North Slope gas. Without this, the Alaskan gas may never be shipped to American consumers.

Adequate supply volume might be arranged in either of two ways. The most economical means may by now be politically impossible. It would involve returning to the rejected concept of moving both Prudhoe Bay and Mackenzie Delta gas through a single pipeline along the shortest route (across the Arctic coastal plain and up the Mackenzie Valley). The Beaufort Sea discoveries assure that such a line could operate at full capacity, sharply reducing cost.

The other alternative would be to defer building the Alaska Highway pipeline for five to ten years, until the late 1980s or early 1990s, in the hope that there would then be enough North Slope gas to fill such a line to capacity. The prospects of this hinge on the discovery of more gas on the North Slope and upon some complex but fascinating factors involved in producing the largest oil and gas reservoir in North America.

The oil and gas at Prudhoe Bay are in the pores of a sandstone rock, lying eight to nine thousand feet below the surface and sloping upward toward the northeast. At the top of the slope, gas is trapped in the pores of the rock, forming the gas cap; below this the sandstone contains a mixture of gas and oil, and still farther down, water is trapped in the sandstone.

In total, some thirty-one billion barrels of oil and thirty-five trillion cubic feet of gas are trapped in the sandstone at Prudhoe Bay, only a portion of which can be brought to the surface. The gas and fluids in the sandstone are under great pressure, which causes them to rise up the bore holes of the producing wells. But as the gas and fluids are produced, the pressure decreases, until finally a substantial portion of the oil and gas is left in the ground with no way to recover it.

The more the reservoir pressure can be maintained, the more oil and gas can be recovered. Along with the oil that is produced comes a mixture of water and gas; these are pumped back down into the sandstone to help minimize the pressure loss. The gas produced with the oil will be re-injected until a pipeline is built to transport it to markets.

As a further means to help maintain reservoir pressure, the oil companies plan to pump water from the Arctic Ocean down injection wells into the reservoir rock. In the final stages of production, water from the ocean will be pumped into the ground at a faster rate than the oil is pumped out, in an operation known as waterflooding. The final plans for waterflooding were to be finalized in 1979. But preliminary studies submitted by the oil companies to the State of Alaska envisage pumping anywhere from eight billion to sixteen billion barrels of ocean water into the reservoir, at rates of up to three and a half million barrels a day. By this means, the oil companies hope to recover forty per cent of the oil in place, or twelve billion barrels, and seventy-five percent of the natural gas, some twenty-six trillion cubic feet.

The production plans approved by the State in 1977 called for the gas produced with the oil to be re-injected for the first five years of oil production, until 1983, when it was hoped that there would be a gas pipeline. The oil companies said they needed at least two years of oil production experience to see how the reservoir actually performs, before they can finalize an optimum means to recover as much of the energy as possible. "Until you have lived with a new wife or lived with a new reservoir, you don't know what is going to be required to keep her happy," is the way Atlantic Richfield vice-president C.O. Goldsmith explained it to Congressional committee hearings in Washington.

In general terms, though, it is clear that a trade-off is involved in selecting the right time to stop re-injecting Prudhoe Bay gas and start producing it for shipment by pipeline. If no gas were produced, in about twenty or thirty years time, it would be necessary to re-inject an estimated fifteen trillion to twenty trillion cubic feet of gas into the reservoir. Not only would this be expensive, but in the process of compression and re-injection, about three per cent of the gas reserves would be consumed as fuel and thus lost. On the other hand, the earlier the gas is produced for sale, the greater will be the volume of water that must be pumped from the ocean into the reservoir to attain maximum oil recovery. This, too, is costly and consumes energy. Preliminary studies by the producers estimated that deferring the start of gas sales from 1983 to 1988 would reduce the amount of water that would have to be pumped into the ground by four billion barrels.

The other factor in the equation, possibly the decisive one, is the rate at which the gas could be produced for sale without adversely affecting the amount of oil that could be recovered. The producers and the State of Alaska have estimated that if gas sales were to start in 1983, it would be safe to produce two billion cubic feet per day. But if this were deferred for five to ten years, until more oil had been recovered, it may be possible to safely produce the gas at a much faster rate, possibly in excess of three billion cubic feet per day.

The implications of this for an Alaska Highway pipeline are clear enough. By deferring completion until the late 1980s or early 1990s there would be a larger daily volume of gas to move from the Prudhoe Bay field, as well as the prospect of more gas from continuing exploration for new fields.

If, as seems certain, the United States will be forced to reassess how and when to utilize its Alaskan gas supply, Canada, too, will have to take another look at its developing energy supplies.

It is a confusing picture. In the early 1970s industrial gas purchasers were unable to contract for all the supply they sought. They were warned that even their existing supplies might be cut back. The Government of Ontario, among others, was preparing plans for allocating supplies in the event of anticipated shortages. As late as July, 1977, the National Energy Board predicted that gas supply in western Canada would be unable to meet Canadian demand plus authorized export volumes beyond 1983, and that Mackenzie Delta gas would then be needed to fill a growing supply gap. But by 1978 it had become clear that there was a significant change—a change that both industry and government had failed to anticipate. In a report to the federal government in late February, 1979, the National Energy Board found that, for the first time in nine years, Canada had gas supplies surplus to its own future needs and the volumes already licensed for export to the United States. Additional volumes of gas export seemed likely to be approved.

Rapid price increases were the key to this changed supply picture. Average wellhead prices for natural gas in Alberta had jumped from 17.4 cents per thousand cubic feet in 1973 to $1.36 in 1978. The effect was to increase supply and curb demand. Large quantities of natural gas which had previously cost more to produce than they could be sold for, were now added to the available supply. The search for new gas supplies was increased greatly. Demand was held in check not only by the higher prices, but also by conservation measures, economic recession, and surplus supplies of fuel oil resulting from excess oil refining capacity in Ontario and eastern Canada. The pattern in the United States was somewhat similar, although there the improved supply picture was not expected to last long.

Despite this improved picture for natural gas, Canada still has need of new supplies of oil and gas to supplement the conventional reserves in western Canada. Even with record exploration activity and expenditures, oil reserves in the decade to 1979 declined by some twenty-five percent. This declining trend of conventional oil reserves in the west is irreversible, and a similar trend for natural gas is sooner or later inevitable. The most optimistic forecast is that by the mid-1980s production from the western provinces will leave a $5 billion annual shortfall in Canada's oil needs.

The best hope for a large new supply region lies in the Beaufort Sea where Canadian-owned Dome Petroleum and its partners were playing out a $400 million exploration drama. The results could confirm an oil and gas potential which may rival, or even exceed, that of western Canada.

The Delta-Beaufort Sea basin covers an area of thirty thousand square miles on land and out into the ocean to a water depth of one thousand feet. Drilling on land and on artificial gravel islands in water depths to fifty feet had resulted in oil and gas discoveries equivalent to one and a half billion barrels of oil by 1978. That is modest compared with reserves equivalent to thirty billion barrels of oil that had been found in western Canada, but it should be remembered that thirty-five thousand exploratory wells had been drilled in the west and only about two hundred in the Delta.

The major prospects, however, lie farther offshore, past the fifty-foot water depth, where Dome and its partners are drilling. Dome's exploration vice-president John Andruik has predicted that the reserves on land and under the shallow water area will be "peanuts" compared with what will be found under deeper water.

The Geological Survey of Canada has estimated the amount of oil and gas which ultimately be found in the Delta-Beaufort Sea Basin at the equivalent of ten to fifty billion barrels of oil. Dome is even more optimistic, estimating the potential at up to one hundred billion barrels, three times the amount found to date in western Canada.

"The Beaufort Sea is one of the most attractive drilling areas in the world for oil and gas potential," according to Dome. A better assessment of that hope should be available by 1980.

The mere existence of enormous deposits of oil and gas means nothing, however, if the cost exceeds the price that users of energy are willing to pay. The existence of the large reserves of oil in the Athabasca oil sands—nearly as large as those in the Middle East—has been known for more than a century, but until very recently the cost of producing this has been more than the oil was worth. Even in 1979, the economics of developing the Athabasca oil sands remained very marginal. Without an economical means of transportation, the same fate faces much of the oil and gas resources of the western Arctic. Devising a way to provide this transportation is a challenge that has yet to be met. It may well be Canada's most important resource challenge in what remains of this century.

Appendix
A Chronology of Some Events

January, 1968
Atlantic Richfield Company discovers North America's largest oil and gas field at Prudhoe Bay on the North Slope of Alaska.

1969
TransCanada Pipelines and two American gas pipeline firms announce preliminary studies of a pipeline to transport Prudhoe Bay gas along the Arctic coast, up the Mackenzie Valley and across Canada to the midwest region of the United States.

June, 1969
Westcoast Transmission Company announces Mountain Pacific planned to transport Prudhoe Bay gas across northern Canada and British Columbia to markets on the U.S. west coast.

January, 1970
The first oil discovery in the Mackenzie Delta is made by Imperial Oil at Atkinson Point; it later turns out to be a small field.

June 27th
Alberta Gas Trunk Line announces a planned pipeline to transport Prudhoe Bay gas, along a route across Alberta, to both midwestern and west coast American markets. AGTL later forms Gas Arctic Systems Study Group with Canadian National Railways and four U.S. gas pipeline firms.

July 15th
Northwest Project Study Group, backed by TransCanada Pipelines, two American gas companies, and three oil companies with gas re-

serves at Prudhoe Bay, announces a twelve million dollar feasibility study of a pipeline to the American midwest. The planned route would largely by-pass Alberta.

January, 1971
The first Mackenzie Delta gas discovery is made by Imperial Oil at Taglu.

June 8th, 1972
Northwest Project Study Group and Gas Arctic System Study Group join forces to become Arctic Gas. This group later expands to twenty-eight participants, but thirteen firms later drop out.

December 4th
El Paso Gas Company announces a nine million dollar study of an alternative plan to pipe Prudhoe Bay gas across Alaska to the Pacific coast where it would be liquefied and transported by tanker ships.

June 27th, 1973
After a year of discussions, following the merger of the two groups, Arctic Gas companies agree on the route of the proposed pipeline which would extend across Alberta.

October 4th
Marshall Crowe, former president of the Canada Development Corporation and member of the management committee of Arctic Gas, is appointed chairman of the National Energy Board.

November 13th
The construction of the Alyeska pipeline to transport Prudhoe Bay oil is finally approved.

December 6th
Prime Minister Trudeau tells the House of Commons that Canada's interest requires early construction of the proposed pipeline to transport Mackenzie Delta and North Slope gas.

March 8th, 1974
The Canadian government defers plans for offshore drilling in the

Beaufort Sea, pending results of a four million dollar environmental study.

March 21st
Arctic Gas files applications with the Canadian and United States governments for a proposed pipeline after completing fifty million dollars in studies. Mr. Justice Thomas R. Berger is appointed to conduct an inquiry and recommend terms and conditions for a pipeline right of way protecting the northern environment and the interests of northern residents.

July 31st
Alberta Gas Trunk Line president Robert Blair announces plans by Foothills Pipe Lines for its "Maple Leaf" project. The planned Mackenzie Valley pipeline would transport only Delta gas in the event that the Arctic Gas proposal fails.

September 13th
Alberta Gas Trunk Line withdraws from Arctic Gas to pursue the Foothills Maple Leaf project along with Westcoast Transmission Company.

September 24th
El Paso files an application with the U.S. Federal Power Commission for competitive trans-Alaska pipeline and LNG tanker project.

March 3rd, 1975
The Mackenzie Valley Pipeline Inquiry begins public hearings in Yellowknife.

March 27th
Foothills files applications for the Maple Leaf pipeline, drawing on Arctic Gas studies.

May 5th
The Federal Power Commission starts hearings on competitive applications by Arctic Gas and El Paso for movement of North Slope gas.

October 27th

The NEB starts hearings on Arctic Gas and Foothills applications. Several organizations raise legal objections to Marshall Crowe's chairing the hearings due to his prior involvement with Arctic Gas; the board refers the matter to the Federal Court.

December 12th

The Federal Court rules there is no legal impediment to Crowe's position as chairman. The Committee for Justice and Liberty later appeals to the Supreme Court.

March 11th, 1976

The Supreme Court rules that Crowe's participation in the hearings raises valid "apprehension of bias." NEB's hearings will have to start over with a new hearing panel.

April 8th

Northwest Pipeline Corporation of Salt Lake City announces it will apply for a pipeline to transport Prudhoe Bay gas along the Alaska Highway.

April 12th

NEB hearings start again with new three-man panel headed by Jack Stabback.

May 5th

Alberta Gas Trunk Line and Westcoast Transmission sign an agreement with Northwest Pipeline to sponsor in Canada the proposed Alcan pipeline along the Alaska Highway.

July 9th

Northwest Pipeline files application for the U.S. portion of the Alcan pipeline with the FPC. The FPC consolidates this with Arctic Gas and El Paso applications.

August 31st

Foothills files application for the Canadian portion of the Alaska Highway pipeline.

September 20th
Dome Petroleum announces the first gas discovery in the deep water area of the Beaufort Sea.

September 30th
The United States Congress passes the Alaska Natural Gas Transportation Act requiring a decision on the transportation of Alaskan gas to be made by the President and Congress rather than by the FPC. The President is required to submit his decision to Congress by September 1, 1977.

October 22nd
President Ford signs the Alaska Natural Gas Transportation Act.

November 12th
The FPC completes hearings on applications by Arctic Gas, El Paso and Northwest.

November 18th
Justice Berger completes the hearings of the Mackenzie Valley Pipeline Inquiry.

February 1st, 1977
The report of FPC hearing examiner, Judge Nahum Litt, says that the Arctic Gas proposal is "vastly superior" to others.

February 10th
Foothills announces it will file a revised application for a larger diameter Alaska Highway pipeline with twenty-five per cent greater design capacity.

February 28th
The revised Alaska Highway plan filed with the NEB and FPC.

April 8th
The FPC staff report says the Arctic Gas proposal is still vastly superior to the revised Alcan proposal.

April 19th
The Alaska Pipeline Inquiry, headed by Kenneth Lysyk, is ap-

pointed to examine the social and economic aspects of the Alaska Highway line. Environmental Assessment Review Panel also appointed.

May 2nd
The FPC report to President Carter recommends a pipeline route across Canada over El Paso's pipeline and tanker proposal. Two commissioners recommend Arctic Gas route, two the Alaska Highway route.

May 9th
The first volume of Berger's report recommends no pipeline across the northern Yukon and no Mackenzie Valley pipeline for ten years.

May 13th
The NEB completes its hearings on pipeline applications. Foothills withdraws its application for the Maple Leaf line.

July 4th
NEB rejects Arctic Gas application and recommends the Alaska Highway line with modifications. The hearing panel envisions a possible future line along the Dempster Highway to connect delta gas with the Alaska Highway pipeline.

July 27th
The report of the Environmental Assessment Review Panel says the proposed Alaska Highway line is environmentally acceptable.

August 1st
The report of the Lysyk Inquiry says construction in the Yukon for the Alaska Highway line should be delayed until August 1, 1981. Arctic Gas dissolves after expenditures of one hundred and fifty million dollars.

August 8th
Following a special two-day debate in Parliament, Prime Minister Trudeau announces that Canada and the United States will hold negotiations on the proposed Alaska Highway line.

September 9th
Prime Minister Trudeau and President Carter announce agreement in principle on the construction of the pipeline.

September 20th
Canada and the United States sign an agreement "on principles applicable to a northern natural gas pipeline." The agreement calls on American gas shippers to subsidize the cost of moving Delta gas through a possible seven hundred and thirty mile Dempster Highway lateral and calls for a review of the size and capacity of the Alaska Highway line.

September 22nd
President Carter's report and decision sent to Congress.

October 30th
Congress approves the Alaska Highway pipeline.

November 18th
Dome Petroleum announces three more gas discoveries in the Beaufort Sea.

January, 1978
Justice Berger produces the second volume of his Mackenzie Valley Pipeline Inquiry report.

February 3rd
Legislation is introduced into the House of Commons to approve the Alaska Highway line and to establish a Northern Pipeline Agency to supervise construction.

April 11th
Following passage by the House of Commons, the Senate ratifies northern pipeline legislation.

Index

259